Ultrafast Spectroscopy

Quantum information and wavepackets

Ultrafast Spectroscopy

Quantum information and wavepackets

Joel Yuen-Zhou
Research Laboratory of Electronics, Massachusetts Institute of Technology

Jacob J Krich
Department of Physics, University of Ottawa

Ivan Kassal
School of Physics and Mathematics, University of Queensland

Allan S Johnson
Department of Physics, Imperial College London

Alán Aspuru-Guzik
Department of Chemistry and Chemical Biology, Harvard University

IOP Publishing, Bristol, UK

ISBN 978-0-750-31062-8 (ebook)
ISBN 978-0-750-31063-5 (print)

DOI 10.1088/978-0-750-31062-8

Version: 20140801

British Library Cataloguing-in-Publication Data
A catalogue record for this book is available from the British Library.

Published by IOP Publishing, wholly owned by The Institute of Physics, London

IOP Publishing, Temple Circus, Temple Way, Bristol, BS1 6HG, UK

US Office: IOP Publishing, Inc., 190 North Independence Mall West, Suite 601, Philadelphia, PA 19106, USA

Contents

Appendices

The authors

Joel Yuen-Zhou

Joel Yuen-Zhou is currently the Robert J Silbey Postdoctoral Fellow in the Center of Excitonics at the Massachusetts Institute of Technology. He received a BSc in Chemistry and Mathematics from the same school in 2007 and a PhD in Chemical Physics from Harvard University in 2012. Starting in July 2015, he will be an assistant professor in the Department of Chemistry and Biochemistry at the University of California San Diego. His research interests are broadly located in the realm of quantum dynamics, specializing in nonlinear spectroscopy, quantum information, time-dependent density functional theory, and topological phases of matter.

Jacob J Krich

Jacob Krich is an assistant professor in the Department of Physics at the University of Ottawa. He received his BA in Physics from Swarthmore College in Pennsylvania, followed by an MMath from Oxford University, where he was a Rhodes Scholar. He received his PhD in theoretical condensed-matter physics from Harvard University. After receiving his PhD, Jacob was a Ziff Fellow of the Harvard University Center for the Environment and a postdoctoral fellow in the Department of Chemistry and Chemical Biology at Harvard. His research focuses on novel pathways to high efficiency photovoltaics and nonlinear spectroscopies of organic systems.

Ivan Kassal

Ivan Kassal is an ARC DECRA Research Fellow in the School of Mathematics and Physics at the University of Queensland. He received his BS in chemistry from Stanford University in 2006 and his PhD in chemical physics from Harvard University in 2010. He has published a number of papers on exciton transport in photosynthesis and on the application of quantum computers to problems in chemistry. His homepage is at http://www.ivankassal.com.

Allan S Johnson

Allan Johnson is a Marie-Curie Early Stage Researcher and NSERC PGS award holder in the Quantum Optics and Laser Science division at Imperial College London. Previously he obtained his undergraduate degree in Physics–Mathematics at the University of Ottawa. Previous to that, he was an actor and set designer. His research interests include nonlinear and nonperturbative quantum dynamics in the presence of strong and ultrafast laser fields.

Alán Aspuru-Guzik

Alán Aspuru-Guzik is a Professor at Harvard University in the Department of Chemistry and Chemical Biology. He received his doctoral degree from the University of California, Berkeley, and was the recipient of the 35 Innovators under 35 by MIT Technology Review for his contributions to the intersection of quantum information and quantum chemistry. In this context, he is interested in how chemical experiments such as ultrafast spectroscopy can be interpreted using ideas from quantum information. More about his work can be found on his home page, http://aspuru.chem.harvard.edu.

Glossary of common terms

2D-ES	Two-dimensional electronic spectrum
$\mathscr{B}$	Bath
χ	Process matrix (equation (1.5))
DEM	Doubly-excited manifold
DS-FD	Double-sided-Feynman diagram
ε_n	Term in n-th pulse proportional to $e^{-i\omega_n t}$ ($\omega_n > 0$, equation (3.4)) In the RWA, it promotes ket (bra) amplitude from lower (higher) to higher (lower) energy states (equation (3.25))
ε_n^*	Term in n-th pulse proportional to $e^{i\omega_n t}$ In the RWA, it promotes (ket) bra amplitude from lower (higher) to higher (lower) energy states (equation (3.25))
ESA	Excited-state absorption
FC	Franck-Condon
GSB	Ground-state bleach
GSR	Ground-state recovery
GSM	Ground-state manifold
$H_0(\mathbf{R})$	Molecular Hamiltonian as a function of nuclear coordinates $\mathbf{R}$ in the absence of pulses (equation (2.4))
LO	Local oscillator
$\boldsymbol{\mu}$	Dipole operator (equation (2.9))
OQS	Open quantum system
PC	Phase cycling
PE	Photon echo
PES	Potential energy surface
PP'	Pump probe
QPT	Quantum process tomography
RWA	Rotating wave approximation (equation (3.25))
$\mathscr{S}$	System
σ	Pulse duration (equation (3.4))
η	Light pulse electric field strength (equation (3.4))
SE	Stimulated emission
SEM	Singly excited manifold
TG	Transient grating
Ω_{ij}^n	Transition amplitude for ket from $\|j\rangle$ to $\|i\rangle$ or for bra from $\|i\rangle$ to $\|j\rangle$ via ε_n. In the RWA, it is significant only if $\omega_{ij} > 0$ (equation (3.20))
$\Omega_{ji}^{\bar{n}}$	Transition amplitude for ket from $\|i\rangle$ to $\|j\rangle$ or for bra from $\|j\rangle$ to $\|i\rangle$ via ε_n^*. In the RWA, it is significant only if $\omega_{ij} > 0$ (equation (3.20))

Introduction

Ultrafast spectroscopy is a powerful tool for studying excited-state processes in physical systems ranging from atoms and molecules in the gas phase to condensed phases such as proteins, membranes, or solids. In a typical spectroscopic experiment, the system of interest is subjected to a series of incoming optical pulses, which trigger nonequilibrium dynamics that are probed by the outgoing light, which constitutes the signal. The adjective 'ultrafast' refers to the timescale of the phenomena of interest, which can span several orders of magnitude, from attoseconds to nanoseconds, and corresponds also to the timescales associated with the duration of the pulses and the separations between them. The information contained in the signal is one of the few windows we have to understand processes occurring at the nanoscale. Hence, the understanding of such techniques is essential to a wide range of modern research in physics, chemistry, and materials science. There is a certain art to the design of a spectroscopic experiment, and experienced practitioners will often come up with creative ways to craft the optical pulses so that the signal reflects the information they are searching for. The theory of ultrafast spectroscopy, however, may be a daunting and confusing exercise for newcomers to the field, as it often involves mastering a formalism that ranges from electromagnetism to time-dependent quantum mechanics, and depending on the system of interest, to solid state physics, open quantum systems, molecular dynamics, and so on.

This text grew out of our efforts to understand ultrafast spectroscopy in a physically intuitive way and from several different perspectives. One such perspective uses concepts from quantum information theory and open quantum systems, which reframe the spectroscopic enterprise as a quantum process tomography (QPT), a protocol for systematically extracting the maximum amount of information possible from a quantum 'black box.' For our purposes, this box corresponds to the excited-state dynamics of the physical system of interest. Spectroscopy as a QPT then becomes an exercise in preparation and detection of quantum states. This persective gives insight into the sometimes convoluted perturbation-theory calculations. This approach also provides limits on the amount of (quantum) information that a particular spectroscopic signal contains. The other perspective is that of wavepacket dynamics, a 'chemical' approach that focuses on tracking the time-dependent quantum state of nuclear degrees of freedom along various electronic potential energy surfaces. The strengths of this perspective are the great physical insights that can be obtained via intuition rooted in classical and semiclassical mechanics, the computational advantages associated with their numerical implementation, and the visualization opportunities associated with their simulation. We have also sought to provide a single source where one could find answers to questions that were supposedly well-known but scattered over the literature.

This book should be accessible to anyone who has taken a full-year graduate-level course in quantum mechanics, and is recommended to beginning researchers—theorists and experimentalists alike—who are interested in quantum dynamics and its experimental observables. We believe that established researchers will also obtain novel physical insights on possibly familiar topics, depending on whether they come from ultrafast spectroscopy, quantum optics, quantum information, or theoretical chemistry, to mention a few possibilities. The text can also serve as a textbook for specialized courses or workshops, or the examples and code can be adapted to be part of problem sets and exercises. A great portion of the material has emerged from our own recent studies, so the topics are timely material for current research. The discussion is sufficiently detailed to allow the reader to design and calculate results for nonlinear optical experiments on toy models (i.e., the coupled dimer), which can be readily generalized to more complex systems that may appear in a research setup. Computational simulations are important for most nontrivial systems, and we provide detailed descriptions and downloadable code to simulate an array of spectroscopic signals.

The structure of the book is as follows. Chapter 1 introduces the quantum information perspective and is devoted to understanding the process matrix in an open quantum system (OQS), i.e., a system interacting with an environment or bath. Chapter 2 introduces the important model systems that we use for examples throughout the book. Chapters 3 and 4 introduce the interaction of light pulses with molecular systems, focusing on how to predict experimental outcomes from common spectroscopic experiments. We use the time-domain wavepacket approach to understand both linear and nonlinear spectroscopies. Chapter 5 shows how well-crafted ultrafast experiments can reconstruct the quantum process matrix for the singly-excited states of model molecular systems. Finally, chapter 6 contains a detailed discussion of the numerical simulation of spectra from a time-dependent perspective and should be useful for readers who are looking to implement computationally the ideas in this book. Example MATLAB® code can be downloaded on the book's website. We conclude with chapter 7 by describing the value and insights of rethinking spectroscopy in terms of quantum information processing concepts. There are six appendices, which address the mathematical description of a short pulse of light, the validity of time-dependent perturbation theory in optical spectroscopy, the equivalence of the many non-interacting molecule calculation and the single-molecule one, a primer on frequency-resolved linear and nonlinear spectroscopy, an introduction to 2D spectroscopy, and finally, the procedure to evaluate the isotropic average of a spectroscopic signal. In particular, appendices A, B, and C offer our perspectives on 'well-known' subjects that are often not covered in elementary introductions. These are our attempts to fill this gap in the literature.

The book is naturally biased in its selection of topics, and is far from a comprehensive treatise for ultrafast spectroscopy (for the latter, we invite the reader to consult the classic textbook by Shaul Mukamel, *Principles of Nonlinear Optical Spectroscopy*, Oxford University Press 1999). First, we restrict ourselves to electronic spectroscopy, although it is clear that most of the tools presented here can be generalized to other energy scales. Second, the material is presented using the model system of the coupled dimer, whereas many real systems are more complicated. This model effectively demonstrates the key ideas of the spectroscopic techniques and can be easily adapted to treat more complex systems. We emphasize the techniques of frequency-integrated pump–probe (PP′) and transient-grating (TG) spectroscopies;

we give brief, self-contained introductions to frequency-resolved and two-dimensional spectroscopies in the appendices. Even though there is presently a wealth of activity in the field of multi-dimensional spectroscopy, these techniques are not significantly different in spirit from their PP′ and TG counterparts, and can be readily understood after having a solid grasp on these techniques. Our aim is to offer what we believe is the simplest introduction to the field that brings the reader up to speed in terms of current research topics, yet without sacrificing physical intuition. In this regard, many interesting subjects such as Raman spectroscopy, response theory, field quantization, optical activity, infrared spectroscopy, semiconductor optics, and applications of 2D spectroscopy have been deliberately omitted. We refer the reader to the book by Mukamel for a survey of such subjects, as well as the texts of Minhaeng Cho (*Two Dimensional Optical Spectroscopy*, CRC Press 2009) as well as Martin Zanni and Peter Hamm (*Concepts and Methods of 2D Infrared Spectroscopy*, Cambridge University Press 2011).

The structure of the book is amenable to a front-to-back reading, but readers with particular interests may easily skip chapters as several of them are self-contained. For example, if one is solely interested in understanding nonlinear spectroscopy via the wavepacket rather than the quantum information approach, chapters 2, 3, and 4 will suffice. Also, chapter 6 can be simply used as a reference while exploring the example MATLAB® code.

A few years ago, we were amongst those confused researchers entering the field of ultrafast spectroscopy. This book compiles many of the insights we gathered throughout the process of doing research on the subject, either by reading textbooks and papers, attending summer schools and conferences, but also speaking with theorists and experimentalists at a personal level. In particular, we wish to acknowledge many discussions with our colleagues Dylan Arias, Keith Nelson, and Semion Saikin. Our approach to the subject is heavily influenced by the teachings of Eric Heller as well as David Tannor's text *Introduction to Quantum Mechanics: A Time-dependent Perspective*, University Science Books, 2007. JYZ and AAG acknowledge support from the US Department of Energy, Office of Science, Office of Basic Energy Sciences under Award Number DESC0001088. JJK and AJ acknowledge support from NSERC. IK was supported by the Australian Research Council, under projects DE140100433, CE110001013, and CE110001027. We are sincerely grateful for the patience and encouragement from IOP editors John Navas and Jacky Mucklow.

With this, we hope that our text will help the reader acquire an appreciation for the inner workings of nonlinear spectroscopy and use those concepts in creative research contexts. We welcome comments and errata at joelyuenzhou@gmail.com.

The authors
Cambridge, Ottawa, Brisbane, London, 2014

IOP Publishing

Ultrafast Spectroscopy
Quantum Information and Wavepackets
Joel Yuen-Zhou *et al*

Chapter 1

The process matrix and how to determine it: quantum process tomography

Our goal in this book is to describe how to extract the maximum possible information about an electronic system even while it interacts with a large environment. The environment (also called the bath) consists of all the degrees of freedom in which we are not interested, such as vibrational motions of the molecules or the solvent. The field of open quantum systems (OQS) describes such problems, in which there is a system of interest and a bath that interacts with the system. We begin by introducing the formalism of the quantum process matrix χ, which is the main mathematical object to reconstruct in QPT. In later chapters, we will see how we can retrieve χ from nonlinear spectroscopy experiments.

We consider an arbitrary system $\mathscr{S}$ interacting with a bath $\mathscr{B}$, such that the total Hamiltonian which governs the two as a whole is,

$$H_0 = H_{\mathscr{S}} + H_{\mathscr{B}} + H_{\mathscr{S}\mathscr{B}}, \tag{1.1}$$

where $H_{\mathscr{S}}, H_{\mathscr{B}}, H_{\mathscr{S}\mathscr{B}}$ depend only on degrees of freedom of $\mathscr{S}$, $\mathscr{B}$, or both, respectively. We are interested in the evolution of $\mathscr{S}$ as a function of time T in the form of a reduced system density matrix $\rho(T) = \mathrm{Tr}_{\mathscr{B}}[\rho_{\mathrm{total}}(T)]$, where the trace is taken over the $\mathscr{B}$ degrees of freedom only. We will show in example 1 that if $\rho_{\mathrm{total}}(0) = \rho(0) \otimes \rho_{\mathscr{B}}(0)$ (i.e., the total initial state has no system–bath correlations), the state $\rho(T)$ of $\mathscr{S}$ at every time $T \geqslant 0$ can be expressed as [5, 23]

$$\rho(T) = \chi(T)\rho(0), \tag{1.2}$$

where the 'superoperator' $\chi(T)$ is the quantum process matrix, a linear operator on the space of density matrices.[1] If the Hilbert space of $\mathscr{S}$ is N-dimensional, $\rho(T)$ is an $N \times N$ matrix and $\chi(T)$ an $N \times N \times N \times N$ tensor. Equation (1.2) indicates that $\chi(T)$ is a linear transformation acting on $\rho(0)$ to produce $\rho(T)$. Since equation (1.2) is true for all initial

[1] We note that, in general, $\chi(T_1 + T_2) \neq \chi(T_2)\chi(T_1)$. Important exceptions are the case of Markovian baths $\mathscr{B}$ and isolated systems [3, 15, 21].

doi:10.1088/978-0-750-31062-8ch1

Figure 1.1. A quantum black box is probed by a series of input states $\rho(0)$ which yield output states corresponding to $\rho(T) = \chi(T)\rho(0)$, where $\chi(T)$ is the process matrix that fully characterizes the black box. The goal of QPT is to reconstruct $\chi(T)$ with a finite number of input–output relations.

states of the system $\rho(0)$, knowledge of $\chi(T)$ consitutes total knowledge of the system and its interaction with the bath, given a fixed $\rho_{\mathscr{B}}(0)$. Equation (1.2) can be regarded as an integrated equation of motion for every T, and in principle, does not make assumptions about the nature of the bath, and therefore holds for both Markovian and non-Markovian dynamics. If $\mathscr{S}$ is an unknown *quantum black box* (figure 1.1), knowledge of $\chi(T)$ completely characterizes it. $\chi(T)$ is known as the process matrix, and its reconstruction via a finite number of experiments is the main goal of QPT.

Equation (1.2) can be expressed in terms of a basis for the system:

$$\rho_{ab}(T) = \sum_{cd} \chi_{abcd}(T)\rho_{cd}(0). \tag{1.3}$$

The interpretation of $\chi(T)$ is easy to grasp: if the initial state is prepared at $\rho(0) = |c\rangle\langle d|$, $\chi_{abcd}(T)$ is the value of the entry *ab* of the density matrix after time T, i.e. $\chi_{abcd}(T) = \langle a|\rho(T)|b\rangle$. Diagonal elements ($\rho_{aa}$) of a density matrix are called *populations*, whereas off-diagonal ones (ρ_{ab} for $a \neq b$) are known as *coherences*[2]. Therefore, $\chi_{abcd}(T)$ denotes a state-to-state amplitude transfer, and indicates a process where a population or a coherence $|a\rangle\langle b|$ at time T gets transferred to another population or coherence $|c\rangle\langle d|$ at time T.

Example 1. Properties of the process matrix $\chi(T)$.

We will now go through the standard derivations of the existence of $\chi(T)$ and of its symmetries. The reader uninterested or already familiar with $\chi(T)$ can skip questions 1–5, and go directly to question 6, where we relate $\chi(T)$ to the Redfield equations. Here and throughout this book we set Planck's constant $\hbar$ equal to 1.

1. Express $\rho_{\text{total}}(T)$ as the time evolution of $\rho_{\text{total}}(0)$ due to H_0. By taking a trace of $\rho_{\text{total}}(T)$ with respect to $\mathscr{B}$, prove that:

$$\rho(T) = \sum_i E_i(T)\rho(0)E_i^\dagger(T). \tag{1.4}$$

[2] Note that $\chi_{abcd}(T)$ and $\rho_{ab}(T)$ depend on the basis.

The terms $E_i(T)$ are called Kraus operators, where i is a multi-index, running over all pairs of $\mathscr{B}$ states. Find an expression for $E_i(T)$ in terms of the evolution operator for the entire $\mathscr{S}$ and $\mathscr{B}$ spaces $U(T) = \mathscr{T}(e^{-i\int_0^T H_0(t')dt'})$, where $\mathscr{T}$ is the time-ordering operator (H_0 could in general be time-dependent).

2. By introducing a basis for $\mathscr{S}$ in equation (1.4), show that $\rho_{ab}(T) = \sum_{cd}\chi_{abcd}(T)\rho_{cd}(0)$.
3. Prove that the expression obtained for $\chi_{abcd}(T)$ can be reexpressed as,

$$\chi_{abcd}(T) = \mathrm{Tr}_B\{\langle a|U(T)(|c\rangle\langle d| \otimes \rho_B(0))U^\dagger(T)|b\rangle\}. \tag{1.5}$$

4. Derive the following symmetries of $\chi(T)$:
   (a) **Hermiticity:** $\chi_{abcd}(T) = \chi^*_{badc}(T)$. Also, show that if $\rho(0)$ is Hermitian, then $\rho(T)$ will be too if this symmetry is satisfied.
   (b) **Trace preservation:** $\sum_a \chi_{aacd}(T) = \delta_{cd}$. Also, show that this condition preserves the trace of $\rho(T)$ at all times.
   (c) **Positivity:** $\sum_{abcd} z^*_{ac}\chi_{abcd}(T)z_{bd} \geqslant 0$ for arbitrary complex matrices z. Also, show that if $\rho(0)$ is positive semidefinite, then this condition guarantees that $\rho(T)$ is too.
   (d) Given the properties in (a)–(c), for a system in d-dimensional Hilbert space, how many linearly independent real-valued elements parametrize $\chi(T)$ completely?
5. Suppose $H_{\mathscr{S}\mathscr{B}} = 0$ so $\mathscr{S}$ is a perfectly isolated system. By expressing $\chi(T)$ in the *eigenbasis* of $H_{\mathscr{S}}$, show that $\chi_{abcd}(T) = \delta_{ac}\delta_{bd}e^{-i\omega_{ab}T}$, where $\omega_{ab} = E_a - E_b$ and δ_{ij} is the Kronecker delta function. Interpret this result as the possible coherence and population transfers that $H_{\mathscr{S}}$ allows.
6. A simple model of Markovian OQS dynamics is given by the secular Redfield equations. Consider a two level system (TLS) $\mathscr{S}$ with eigenstates $|a\rangle$ and $|b\rangle$, which from coupling to $\mathscr{B}$ exhibits the following kinetics,

$$\dot{\rho}_{aa}(T) = -k_{b\leftarrow a}\rho_{aa}(T), \tag{1.6a}$$
$$\dot{\rho}_{bb}(T) = k_{b\leftarrow a}\rho_{aa}(T), \tag{1.6b}$$
$$\dot{\rho}_{ab}(T) = -i\omega_{ab}\rho_{ab}(T) - \kappa_{ab}\rho_{ab}(T), \tag{1.6c}$$
$$\dot{\rho}_{ba}(T) = [\dot{\rho}_{ab}(T)]^*. \tag{1.6d}$$

Derive analytical expressions for $\chi_{abcd}(T)$ for this particular model and interpret their physical significance.

Solution

1. Assume that the initial state of ρ_{total} is in a tensor product form,

$$\rho_{\text{total}}(0) = \rho(0) \otimes \rho_{\mathscr{B}}(0), \tag{1.7}$$

where $\rho_{\mathscr{B}}(0)$ can be written in diagonal form as

$$\rho_{\mathscr{B}}(0) = \sum_\beta p_\beta|e_\beta\rangle\langle e_\beta|, \tag{1.8}$$

for every initial state $\rho(0)$ of the system, with $p_\beta \geqslant 0$ and $\sum_\beta p_\beta = 1$. At time T, the total state is a unitary evolution of the initial total state,

$$\rho_{\text{total}}(T) = U(T)[\rho(0) \otimes \rho_{\mathscr{B}}(0)]U^\dagger(T). \tag{1.9}$$

Taking the trace of equation (1.9) with respect to the states of $\mathscr{B}$ yields,

$$\rho(T) = \sum_{\alpha\beta} p_\beta E_{\alpha\beta}(T)\rho(0)E^\dagger_{\alpha\beta}(T), \tag{1.10}$$

where

$$E_{\alpha\beta}(T) = \langle e_\alpha|U(T)|e_\beta\rangle, \tag{1.11}$$

is a Kraus operator and equation (1.10) is known as the *operator sum representation* [5, 23]. The $\{\alpha, \beta\}$ indexing can be trivially relabeled using a single index i to yield equation (1.4).

2. Sandwiching equation (1.10) between $\langle a|$ and $|b\rangle$ results in:

$$\langle a|\rho(T)|b\rangle = \sum_{cd}\sum_{\alpha\beta} p_\beta \langle a|E_{\alpha\beta}(T)|c\rangle\langle c|\rho(0)|d\rangle\langle d|E^\dagger_{\alpha\beta}(T)|b\rangle. \tag{1.12}$$

Since $\langle a|\rho(T)|b\rangle = \rho_{ab}(T)$, we can immediately write

$$\rho_{ab}(T) = \sum_{cd} \chi_{abcd}(T)\rho_{cd}(0), \tag{1.13}$$

$$\begin{aligned}\chi_{abcd}(T) &= \sum_{\alpha\beta}[E_{\alpha\beta}(t)]_{ac}[E^\dagger_{\alpha\beta}(t)]_{db} \\ &= \sum_{\alpha\beta} p_\beta\langle e_\alpha, a|U(T)|e_\beta, c\rangle\langle e_\beta, d|U^+(T)|e_\alpha, b\rangle. \end{aligned} \tag{1.14}$$

We have then proven the equivalence between equations (1.4) and (1.3) and (1.2).

3. By using equation (1.8), and noticing that the summation over α is a trace over the bath,

$$\begin{aligned}\chi_{abcd}(T) &= \sum_\alpha \langle e_\alpha, a|U(T)[|c\rangle\langle d| \otimes \rho_{\mathscr{B}}(0)]U^\dagger(T)|e_\alpha, b\rangle \\ &= \text{Tr}_{\mathscr{B}}\{\langle a|U(T)[|c\rangle\langle d| \otimes \rho_{\mathscr{B}}(0)]U^\dagger(T)|b\rangle\}. \end{aligned} \tag{1.15}$$

Equation (1.15) indicates that $\chi_{abcd}(T)$ is the amplitude of transfer from $|c\rangle\langle d|$ to $|a\rangle\langle b|$ at time T given the fixed initial state of the bath, $\rho_{\mathscr{B}}(0)$.

4. Hermiticity, trace preservation and positivity properties of $\chi(T)$.
   (a) Manipulating equation (1.14), it follows that

$$\begin{aligned}\chi_{badc}(T) &= \sum_{\alpha\beta} p_\beta\langle e_\alpha, b|U(T)|e_\beta, d\rangle\langle e_\beta, c|U^\dagger(T)|e_\alpha, a\rangle \\ &= \left(\sum_{\alpha\beta} p_\beta\langle e_\alpha, a|U(T)|e_\beta, c\rangle\langle e_\beta, d|U^\dagger(T)|e_\alpha, b\rangle\right)^* \\ &= \chi^*_{abcd}(T). \end{aligned} \tag{1.16}$$

Equation (1.16) preserves the Hermiticity of the density matrix:

$$\begin{aligned}
\rho_{ba}(T) &= \sum_{cd} \chi_{badc}(T)\rho_{dc}(0) \\
&= \sum_{cd} \chi^*_{abcd}(T)\rho^*_{cd}(0) \\
&= \left[\sum_{cd} \chi_{abcd}(T)\rho_{cd}(0)\right]^* \\
&= \rho^*_{ab}(T). \qquad (1.17)
\end{aligned}$$

(b) Using equation (1.14), and exploiting the fact that $U(T)U^\dagger(T) = U^\dagger(T)U(T) = \mathbb{I} \equiv \mathbb{I}_\mathscr{S} \otimes \mathbb{I}_\mathscr{B}$, the identity on the whole space, we get:

$$\begin{aligned}
\sum_a \chi_{aacd}(T) &= \sum_{a\alpha\beta} p_\beta \langle e_\alpha, a|U(T)|e_\beta, c\rangle\langle e_\beta, d|U^\dagger(T)|e_\alpha, a\rangle \\
&= \sum_{a\alpha\beta} p_\beta \langle e_\beta, d|U^\dagger(T)|e_\alpha, a\rangle\langle e_\alpha, a|U(T)|e_\beta, c\rangle \\
&= \sum_\beta p_\beta \langle e_\beta, d|e_\beta, c\rangle \\
&= \delta_{cd}. \qquad (1.18)
\end{aligned}$$

Equation (1.18) preserves the trace of the density matrix as a function of T:

$$\begin{aligned}
\mathrm{Tr}(\rho(T)) &= \sum_{acd} \chi_{aacd}(T)\rho_{cd}(0) \\
&= \sum_{cd} \delta_{cd}\rho_{cd}(0) \\
&= \mathrm{Tr}(\rho(0)). \qquad (1.19)
\end{aligned}$$

(c) Again, manipulating equation (1.14), for any $z_{ij} \in \mathbb{C}$:

$$\begin{aligned}
\sum_{abcd} z^*_{ac}\chi_{abcd}(T)z_{bd} &= \sum_{abcd} z^*_{ac} \sum_{\alpha\beta} p_\beta \langle e_\alpha, a|U(T)|e_\beta, c\rangle \\
&\quad \langle e_\beta, d|U^\dagger(T)|e_\alpha, b\rangle z_{bd} \\
&= \sum_{\alpha\beta} p_\beta \zeta_{\alpha\beta}\zeta^*_{\alpha\beta} \\
&\geqslant 0, \qquad (1.20)
\end{aligned}$$

which is the definition of positivity, and where we have defined $\zeta_{\alpha\beta} = \sum_{ac} z^*_{ac}\langle e_\alpha, a|U(T)|e_\beta, c\rangle$.

Suppose $\rho(0)$ is Hermitian positive semidefinite, so that we may write $\rho_{cd}(0) = \sum_k V^*_{kc} q_k V_{kd}$ where V is a unitary transformation that diagonalizes

$\rho(0)$, and $q_k \geqslant 0$ for all k. Is the Hermitian positive-semidefinite condition maintained for $\rho(T)$? If so, it must satisfy $\sum_{ab} y_a^* \rho_{ab}(T) y_b \geqslant 0$ for an arbitrary vector y. Equation (1.20) guarantees this:

$$\begin{aligned}\sum_{ab} y_a^* \rho_{ab}(T) y_b &= \sum_{abcdk} y_a^* \chi_{abcd}(T) V_{kc}^* q_k V_{kd} y_b \\ &= \sum_k \sum_{abcd} z_{ac}^{(k)*} \chi_{abcd}(T) z_{bd}^{(k)} \\ &\geqslant 0,\end{aligned} \tag{1.21}$$

where we have identified the vectors $z^{(k)}$ with elements $z_{bd}^{(k)} = \sqrt{q_k} V_{kd} y_b$.

(d) Without the constraints of parts (a)–(c), there are $2d^4$ real-valued elements parametrizing a $\chi(T)$ matrix (a real and an imaginary part for each of the d^4 entries).

- Let us now consider the constraints due to Hermiticity (equation (1.16)). We distinguish two cases. First, there are $2d^2$ parameters for population transfer terms $\chi_{aabb}(T)$, but the constraint $\Im\chi_{aabb}(T) = 0$ removes half of them, leaving d^2 parameters. Second, there are the remaining terms $\chi_{abcd}(T)$ where $a \neq b$ or $c \neq d$. There are $2d^4 - 2d^2$ such terms, which denote population-to-coherence, coherence-to-population, and coherence-to-coherence processes, and they are constrained by,

$$\Re\chi_{abcd}(T) = \Re\chi_{badc}(T), \tag{1.22a}$$

$$\Im\chi_{abcd}(T) = -\Im\chi_{badc}(T), \tag{1.22b}$$

  which again, reduces the number of free parameters by half, $d^4 - d^2$. So at the end, from Hermiticity we are left with $2d^4 - d^2 - (d^4 - d^2) = d^4$ parameters.

- Trace preservation (equation (1.18)) imposes additional constraints. We treat the two cases separately again. If the initial state is a population, we have d possibilities for $|c\rangle\langle c|$, which gives d equations of the form $\sum_a \chi_{aacc}(T) = 1$. If the initial state is a coherence $|c\rangle\langle d|$ for $c \neq d$, we have $\frac{d(d-1)}{2}$ different pairings ($|d\rangle\langle c|$ is the same since we already exploited Hermiticity), each giving two constraints, $\Re\sum_a \chi_{aacd}(T) = 0$ and $\Im\sum_a \chi_{aacd}(T) = 0$, which totals $d(d-1)$ constraints. Altogether, trace preservation subtracts $d + d(d-1) = d^2$ parameters.

  Hence, from Hermiticity and trace preservation, the $2d^4$ free parameters of $\chi(T)$ end up reducing to $2d^4 - d^4 - d^2 = d^4 - d^2$ parameters. Additionally, these $d^4 - d^2$ parameters need to satisfy the Hermitian positive-semidefinite condition of equation (1.20).

5. If $H_{\mathcal{SB}} = 0$, then we have an isolated system. Since $\rho(T) = e^{iH_{\mathcal{S}}t}\rho(0)e^{-iH_{\mathcal{S}}t}$, then $\rho_{ab}(T) = \rho_{ab}(0)e^{-i\omega_{ab}T}$, and it immediately follows that $\chi_{abcd}(T) = \delta_{ac}\delta_{bd}e^{-i\omega_{ab}T}$. The result is rather simple and can be interpreted as follows: a population in the eigenbasis of $|c\rangle\langle c|$ remains $|c\rangle\langle c|$, but a coherence $|c\rangle\langle d|$ evolves as $e^{-i\omega_{cd}T}|c\rangle\langle d|$, i.e., it simply picks up a phase factor depending on the difference in energy between the two states under consideration.

6. Similarly, integrating equations (1.6*a*) (1.6*d*), we immediately have the following nonzero elements of $\chi(T)$:

$$\chi_{aaaa}(T) = \mathrm{e}^{-k_{b\leftarrow a}T}, \tag{1.23a}$$

$$\chi_{bbaa}(T) = 1 - \mathrm{e}^{-k_{b\leftarrow a}T}, \tag{1.23b}$$

$$\chi_{bbbb}(T) = 1, \tag{1.23c}$$

$$\chi_{aabb}(T) = 0, \tag{1.23d}$$

$$\chi_{abab}(T) = \mathrm{e}^{-\mathrm{i}\omega_{ab}T}\mathrm{e}^{-\kappa_{ab}T}, \tag{1.23e}$$

$$\chi_{abab}(T) = [\chi_{baba}(T)]^{*}. \tag{1.23f}$$

Here, $\chi_{aaaa}(T)$ and $\chi_{bbaa}(T) = 1 - \chi_{aaaa}(T)$ indicate population decay and transfer from state a with a first order kinetics rate given by $k_{b\leftarrow a}$. Population transfer from $|b\rangle$ to $|a\rangle$ does not occur in this model and therefore if the system is prepared at $|b\rangle$ it will remain there indefinitely. Finally, coherences oscillate at the ω_{ab} frequency but they are damped at a dephasing rate κ_{ab}. No transfers from population to coherence, coherence to population, or coherence to coherence are included in this model.

A possible algorithm to perform QPT is the following: (a) Prepare a linearly independent set of states $\rho(0)$ that spans the vector space of the possible initial density matrices of $\mathcal{S}$; (b) for each of the prepared states, wait for a free evolution time T and determine the density matrix $\rho(T)$. Any protocol for determining a density matrix for a system is called quantum state tomography (QST) [7, 8, 10, 12–14]. In essence, QPT can be carried out for any system with both selective preparation of initial states and QST. Variants of this methodology exist, although all of them operate within the same spirit [6, 17, 18, 20]. QPT has been successfully implemented in a wide variety of experimental scenarios, including nuclear magnetic resonance [4, 11, 24], ion traps [22], single photons [1, 16], solid state qubits [9], optical lattices [19], and Josephson junctions [2]. In this book, we show how to perform QPT for the excited-state dynamics of a simple model multichromophoric system described in chapter 2 using methods of ultrafast spectroscopy. QPT is achieved by controlling the polarization and frequencies of the pulses, which selectively prepare and measure states in the chromophores. This procedure is described in chapter 5; first we will introduce the systems and an analysis of some standard optical spectroscopies.

Bibliography

[1] Altepeter J B, Branning D, Jeffrey E, Wei T C, Kwiat P G, Thew R T, O'Brien J L, Nielsen M A and White A G 2003 Ancilla-assisted quantum process tomography *Phys. Rev. Lett.* **90** 193601

[2] Bialczak R C, Ansmann M, Hofheinz M, Lucero E, Neeley M, O'Connell A D, Sank D, Wang H, Wenner J, Steffen M, Cleland A N and Martinis J M 2010 Quantum process tomography of a universal entangling gate implemented with Josephson phase qubits *Nat. Phys.* **6** 409–13

[3] Breuer H-P and Petruccione F 2002 *The Theory of Open Quantum Systems* (New York: Oxford University Press)

[4] Childs A M, Chuang I L and Leung D W 2001 Realization of quantum process tomography in NMR *Phys. Rev.* A **64** 012314

[5] Choi M D 1975 Completely positive linear maps on complex matrices *Linear Algebra Appl.* **10** 285–90

[6] Chuang I L and Nielsen M A 1997 Prescription for experimental determination of the dynamics of a quantum black box *J. Mod. Opt.* **44** 2455–67

[7] Cina J A 2000 Nonlinear wavepacket interferometry for polyatomic molecules *J. Chem. Phys.* **113** 9488–96

[8] Dunn T J, Walmsley I A and Mukamel S 1995 Experimental determination of the quantum-mechanical state of a molecular vibrational mode using fluorescence tomography *Phys. Rev. Lett.* **74** 884–7

[9] Howard M, Twamley J, Wittmann C, Gaebel T, Jelezko F and Wrachtrup J 2006 Quantum process tomography and Linblad estimation of a solid-state qubit *New J. Phys.* **8** 33

[10] Humble T S and Cina J A 2004 Molecular state reconstruction by nonlinear wave packet interferometry *Phys. Rev. Lett.* **93** 060402

[11] Kampermann H and Veeman W S 2005 Characterization of quantum algorithms by quantum process tomography using quadrupolar spins in solid-state nuclear magnetic resonance *J. Chem. Phys.* **122** 214108

[12] Leichtle C, Schleich W P, Averbukh I Sh and Shapiro M 1998 Quantum state holography *Phys. Rev. Lett.* **80** 1418–21

[13] Leonhardt U 1995 Quantum-state tomography and discrete Wigner function *Phys. Rev. Lett.* **74** 4101–5

[14] Loh Z H, Khalil M, Correa R E, Santra R, Buth C and Leone S R 2007 Quantum state-resolved probing of strong-field-ionized xenon atoms using femtosecond high-order harmonic transient absorption spectroscopy *Phys. Rev. Lett.* **98** 143601

[15] May V and Kuhn O 2004 *Charge and Energy Transfer Dynamics in Molecular Systems* (New York: Wiley-VCH)

[16] Mitchell M W, Ellenor C W, Schneider S and Steinberg A M 2003 Diagnosis, prescription, and prognosis of a Bell-state filter by quantum process tomography *Phys. Rev. Lett.* **91** 120402

[17] Mohseni M and Lidar D A 2006 Direct characterization of quantum dynamics *Phys. Rev. Lett.* **97** 170501

[18] Mohseni M, Rezakhani A T and Lidar D A 2008 Quantum-process tomography: resource analysis of different strategies *Phys. Rev.* A **77** 032322

[19] Myrskog S H, Fox J K, Mitchell M W and Steinberg A M 2005 Quantum process tomography on vibrational states of atoms in an optical lattice *Phys. Rev.* A **72** 013615

[20] Nielsen M A and Chuang I L 2000 *Quantum Computation and Quantum Information* (Cambridge: Cambridge University Press)

[21] Nitzan A 2006 *Chemical Dynamics in Condensed Phases* (Oxford: Oxford University Press)

[22] Riebe M, Kim K, Schindler P, Monz T, Schmidt P O, Körber T K, Hänsel W, Häffner H, Roos C F and Blatt R 2006 Process tomography of ion trap quantum gates *Phys. Rev. Lett.* **97** 220407

[23] Sudarshan E C G, Mathews P M and Jayaseetha R 1961 Stochastic dynamics of quantum-mechanical systems *Phys. Rev.* **121** 920–4

[24] Weinstein Y S, Havel T F, Emerson J, Boulant N, Saraceno M, Lloyd S and Cory D G 2004 Quantum process tomography of the quantum Fourier transform *J. Chem. Phys.* **121** 6117–33

Chapter 2

Model systems and energy scales

We will now introduce prototypical model systems in order to develop our intuition about QPT and spectroscopy. We will consider the electronic space to be the system and all vibrations, both in the molecules and surroundings, to be the bath. The idea is to consider molecules described by the energy diagram depicted in figure 2.1, which can be considered a listing of eigenstates of H_0, including both the system $\mathscr{S}$ and the bath $\mathscr{B}$. At the beginning $T = 0$, the molecule is in some state in the ground-state manifold (GSM) consisting of the molecular electronic ground state and some particular vibrational eigenstate of $\mathscr{B}$ (or an incoherent mixture of such vibrational eigenstates, which can be analysed independently). Upon resonant interaction with light, which we will consider to be in the visible or ultraviolet frequency, amplitude from this state can be transferred to the singly-excited manifold (SEM). Nonequilibrium dynamics ensue. In order to monitor these dynamics, another pulse can be applied, which transfers amplitude from the SEM to the doubly-excited manifold (DEM), or alternatively, via stimulated emission back to the GSM. This process is the essence of a host of techniques under the broad umbrella of pump–probe (PP') spectroscopy, in order to get a deep understanding of this spectroscopy, it is instructive to develop theoretical tools such as careful book-keeping of energy loss or gain by different pulses passing through the sample. These ideas will be thoroughly explained in chapters 3 and 4.

We shall focus on characteristic energy scales in organic dye molecules that absorb light in the visible part of the electromagnetic spectrum. For simplicity, we will assume that transitions between the GSM and SEM, and between the SEM and DEM, are dipole allowed, but not from the GSM to the DEM. We will further assume that the optical frequencies are not resonant with transitions between states of the same manifold, which can thus be ignored. The typical separation between vibronic eigenstates in each manifold is at most 3600 cm^{-1}, but the separation between manifolds is on the order of optical frequencies, 12 000–25 000 cm^{-1}, which is the range of center frequencies of the pulses. Note that because of this separation of energy scales, an initial thermal distribution of molecules will have considerable

doi:10.1088/978-0-750-31062-8ch2

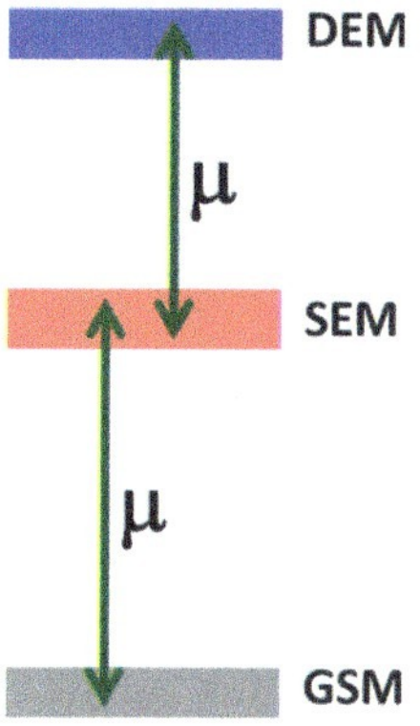

Figure 2.1. Energy level diagram of a model molecular system. It consists of a ground-state manifold (GSM), singly-excited manifold (SEM) and doubly-excited manifold (DEM). Dipole allowed optical transitions are indicated by green arrows.

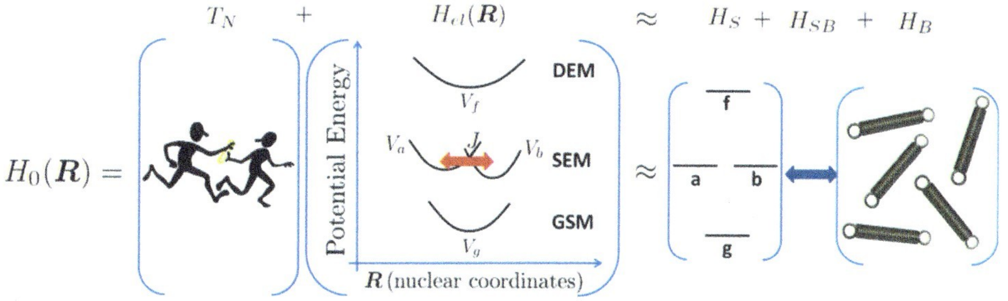

Figure 2.2. Partition of molecular Hamiltonian $H_0(\mathbf{R})$ in terms of potential energy surfaces (PES) (left) or approximation as a sum of an electronic system, a harmonic oscillator (HO) bath, and a coupling between the two (right). This is an approximation because, in general, molecular PES are anharmonic.

population only in the GSM, but not in the SEM or the DEM. We typically consider Gaussian pulses with an optical central frequency and a bandwidth of about 500 cm^{-1}, corresponding to a duration of a few tens of femtoseconds. As we shall see, the dynamics of interest occurs on the timescale of hundreds of femtoseconds to picoseconds, so one may effectively regard the pulses as short compared to the dynamics. Hence, the physical picture that we shall derive is that a series of pulses acts impulsively (i.e., almost instantaneously) to prepare and detect amplitudes, whereas the dynamics of interest is generated by H_0, the Hamiltonian in the absence of pulses. Additionally, for pedagogical purposes, we shall ignore nonradiative pathways between the different manifolds in this book. Spontaneous emission is safely ignored since it generally occurs on a timescale of nanoseconds.

To be more explicit, we consider a toy model, the *coupled dimer* (see figure 2.2) [1–4], constructed by coupling two two-level molecules (sites) a and b. Its Hamiltonian is given by

$$\boxed{H_0(\mathbf{R}) = T_N + H_{el}(\mathbf{R}), \qquad (2.1)}$$

where $\mathbf{R} = (x_1, \ldots, x_n)$ denotes the nuclear coordinates, $T_N = \sum_i p_i^2/2m_i$ is the nuclear kinetic energy, and $H_{el}(\mathbf{R})$ is given by

$$H_{el}(\mathbf{R}) = \sum_{i=g,a,b,f} V_i(\mathbf{R})|i\rangle\langle i| + J(\mathbf{R})(|a\rangle\langle b| + |b\rangle\langle a|). \tag{2.2}$$

$V_i(\mathbf{R})$ are the diabatic potential energy surfaces (PES)[1], and $J(\mathbf{R})$ is the coupling between the two site-excitations $|a\rangle$ and $|b\rangle$ in the SEM.

Let us consider two different partitions of $H_0(\mathbf{R})$, which will be useful in the following chapters.

1. **Excitation manifolds.** Defining the nuclear Hamiltonian for each surface,

$$H_i(\mathbf{R}) = [T_N + V_i(\mathbf{R})]|i\rangle\langle i|, \tag{2.3}$$

we may partition $H_0(\mathbf{R})$ in equation (2.1) into a sum of contributions acting separately on different excitation manifolds,

$$H_0(\mathbf{R}) = H_{\mathrm{GSM}}(\mathbf{R}) + H_{\mathrm{SEM}}(\mathbf{R}) + H_{\mathrm{DEM}}(\mathbf{R}), \tag{2.4}$$

$$H_{\mathrm{GSM}}(\mathbf{R}) = H_g(\mathbf{R}), \tag{2.5}$$

$$H_{\mathrm{SEM}}(\mathbf{R}) = H_a + H_b + J(\mathbf{R})(|a\rangle\langle b| + |b\rangle\langle a|), \tag{2.6}$$

$$H_{\mathrm{DEM}}(\mathbf{R}) = H_f(\mathbf{R}). \tag{2.7}$$

In words, the GSM has only one electronic state $|g\rangle$, the SEM has two states $|a\rangle$ and $|b\rangle$, and the DEM has again only one state $|f\rangle$. The partition given in equation (2.4) is useful if one is interested in the explicit evolution of the different nuclear wavepackets, which we will consider in chapter 4. Physically, we are considering two chromophores (sites), each with only two electronic states, a ground and an excited state. $|g\rangle$ is the electronic state where both chromophores are in their ground states. $|a\rangle$ and $|b\rangle$ are the electronic states where only one chromophore site (a or b, respectively) is excited. Yet, via the Coulomb interaction $J(\mathbf{R})$, the excitation in a can be passed onto b (or the other way around). $|f\rangle$ is the electronic state where both a and b are excited (see figure 2.2). Notice that we are ignoring coupling terms between the excitation manifolds, and, therefore, we also disregard the possibility of conical intersections between the latter, although not within the SEM itself. Inclusion of substantial nonadiabatic dynamics between excitation manifolds is beyond the scope of this book, but is an interesting research topic on its own [8, 9].

[1] See chapter 12 of [7] for a discussion of adiabatic and diabatic PES. The truncation of $H_0(\mathbf{R})$ to a small set of diabatic PES is an approximation to the true molecular Hamiltonian.

2. **Electronic system $\mathscr{S}$ and vibrational bath $\mathscr{B}$.** Alternatively, it is often convenient to rewrite equation (2.1) in the language of OQS, equation (1.1), where the electrons are the system $\mathscr{S}$ and the nuclei are the bath $\mathscr{B}$ (see figure 2.2),

$$H_0(\mathbf{R}) = H_{\mathscr{S}} + H_{\mathscr{B}}(\mathbf{R}) + H_{\mathscr{S}\mathscr{B}}(\mathbf{R}) \tag{2.8a}$$

$$H_{\mathscr{S}} = H_{el}(\mathbf{0}), \tag{2.8b}$$

$$H_{\mathscr{B}}(\mathbf{R}) = T_N + \sum_i \frac{m_i \omega_i^2 x_i^2}{2}, \tag{2.8c}$$

$$H_{\mathscr{S}\mathscr{B}}(\mathbf{R}) = H_{el}(\mathbf{R}) - H_{el}(\mathbf{0}) - \sum_i \frac{m_i \omega_i^2 x_i^2}{2}. \tag{2.8d}$$

Here, $\mathscr{B}$ has been chosen to be a set of harmonic oscillators (HO) with normal-mode coordinates x_i and frequencies ω_i, which amounts to a choice of the form of $V_i(\mathbf{R})$ in equation (2.3). The choice of a HO bath, instead of say, an anharmonic one, is convenient as analytical results associated with HOs are readily available. The frequencies ω_i can be chosen in a physically motivated way (see example 2), such as corresponding to a quadratic expansion of $V_g(\mathbf{R})$ about its global minimum, or such that $H_{\mathscr{S}\mathscr{B}}(\mathbf{R})$ is small in some metric and can be treated as a perturbation on $H_{\mathscr{S}} + H_{\mathscr{B}}$. Note that the right hand sides of equations (2.8*b*)–(2.8*d*) add up to $H_0(\mathbf{R})$. As one can imagine, depending on the physics of the problem, the suggested partition is not unique, and it is often nontrivial to draw the border between $\mathscr{S}$ and $\mathscr{B}$ in a way that $H_{\mathscr{S}\mathscr{B}}$ is manageable. This is an important problem in the theory of OQS, which we shall not address here. We refer the interested reader to the appropriate sources [5, 6].

We will consider transitions between the different vibronic states of the coupled dimer due to coupling of a time-dependent electric field with the dipole operator $\boldsymbol{\mu}$,

$$\boldsymbol{\mu} = \sum_{p=a,b} \left(\boldsymbol{\mu}_{pg}|p\rangle\langle g| + \boldsymbol{\mu}_{fp}|f\rangle\langle p| + \text{h.c.}\right), \tag{2.9}$$

where h.c. indicates Hermitian conjugate. As mentioned at the beginning of the chapter, $\boldsymbol{\mu}$ only allows transitions between different excitation manifolds, and we neglect static dipoles $\boldsymbol{\mu}_{ii}$, which are irrelevant to transitions induced by weak intensity visible–UV light. Importantly, throughout this book, we will assume the Condon approximation, namely, that the dipole transitions between different electronic states $\boldsymbol{\mu}_{ij}$ do not depend on $\mathbf{R}$. In terms of OQS, this means $\boldsymbol{\mu}$, which only acts on $\mathscr{S}$, does not depend on the state of $\mathscr{B}$. Also, we will assume that the values of $\boldsymbol{\mu}_{ij}$ are known.

Example 2. Partition of a molecular Hamiltonian $H_0(\mathbf{R})$ as an open quantum system.

Let us motivate the partition from equations (2.8*b*)–(2.8*d*) a bit further. A quadratic expansion of the PES $V_i(\mathbf{R})$ about their minima yields[2]

$$V_g(\mathbf{R}) = \omega_g + \sum_{i=1}^{N_a+N_b} \frac{m_i\omega_i^2 x_i^2}{2}, \tag{2.10}$$

$$V_a(\mathbf{R}) = \omega_a + \sum_{i=1}^{N_a} \frac{m_i\omega_i^2 (x_i - \Delta_i)^2}{2} + \sum_{i=N_a+1}^{N_a+N_b} \frac{m_i\omega_i^2 x_i^2}{2}, \tag{2.11}$$

$$V_b(\mathbf{R}) = \omega_b + \sum_{i=1}^{N_a} \frac{m_i\omega_i^2 x_i^2}{2} + \sum_{i=N_a+1}^{N_a+N_b} \frac{m_i\omega_i^2 (x_i - \Delta_i)^2}{2}, \tag{2.12}$$

$$V_f(\mathbf{R}) = \omega_a + \omega_b + \sum_{i=1}^{N_a+N_b} \frac{m_i\omega_i^2 (x_i - \Delta_i)^2}{2}, \tag{2.13}$$

$$J(\mathbf{R}) = J. \tag{2.14}$$

These PES represent a very simple model for a coupled dimer. The idea is that the nuclear coordinates from $i = 1$ to $i = N_a$ are localized in molecule a, whereas the ones from $i = N_a + 1$ to $i = N_a + N_b$ are in molecule b. The potential energy surfaces are quadratic with respect to all the nuclei, and there is a single frequency ω_i associated with each nuclear coordinate (which holds for every electronic state). In the ground state, the potential energy minimum is located at $\mathbf{R} = \mathbf{0}$. The displacements Δ_i indicate changes in the equilibrium geometry of the sites (in the absence of coupling J) upon excitation, so that $V_a(\mathbf{R})$ only exhibits these changes for the modes of molecule a, and similarly for $V_b(\mathbf{R})$. Hence, the minima of $V_a(\mathbf{R})$ and $V_b(\mathbf{R})$ are at $\mathbf{R} = (\Delta_1, \ldots, \Delta_{N_a}, 0, \ldots, 0)$ and $\mathbf{R} = (0, \ldots, 0, \Delta_{N_a+1}, \ldots, \Delta_{N_a+N_b})$, respectively. We define the energy minimum of each of these potentials to be at ω_a and ω_b, respectively. V_f describes the potential energy when both molecules are excited, so that the potential energy minimum is at $\omega_a + \omega_b$ and $\mathbf{R} = (\Delta_1, \ldots, \Delta_{N_a+N_b})$. This is an approximation that ignores the interaction between the two excitations (sometimes called binding energy), and so considers that $|f\rangle$ is simply a product state of two excitations, one in each molecule. In this approximation, the following two identities hold,

$$\boldsymbol{\mu}_{ag} = \boldsymbol{\mu}_{fb}, \tag{2.15}$$

$$\boldsymbol{\mu}_{bg} = \boldsymbol{\mu}_{fa}, \tag{2.16}$$

that is, the transition from $|b\rangle$ or $|a\rangle$ to $|f\rangle$ has the same amplitude as the excitation of $|a\rangle$ or $|b\rangle$ from $|g\rangle$, respectively. Since $H_0(\mathbf{R})$ is time-reversal invariant, we may choose the dipole matrix elements as well as the coupling J to be real-valued. For simplicity, we also assume that J is independent of $\mathbf{R}$.

Substitute equations (2.10)–(2.14) into equations (2.8*b*)–(2.8*d*) and interpret the resulting expressions. See figure 2.2.

[2] Here ω_i, where i are integers, are oscillator frequencies, but ω_g, ω_a, and ω_b are base energies of the different PES, with $(\omega_a - \omega_g), (\omega_b - \omega_g) \gg \omega_i$.

Solution

The Hamiltonian for the electronic system $\mathscr{S}$ is given by,

$$\begin{aligned} H_{\mathscr{S}} &= \omega_g |g\rangle\langle g| + (\omega_a + \Lambda_a)|a\rangle\langle a| \\ &+ (\omega_b + \Lambda_b)|b\rangle\langle b| + J(|a\rangle\langle b| + |b\rangle\langle a|) \\ &+ (\omega_f + \Lambda_f)|f\rangle\langle f|, \end{aligned} \tag{2.17}$$

where $\Lambda_a = \sum_{i=1}^{N_a} \frac{m_1 \omega_i^2 \Delta_i^2}{2}$ and $\Lambda_b = \sum_{i=N_a+1}^{N_b} \frac{m_1 \omega_i^2 \Delta_i^2}{2}$ are the *reorganization energies* in the absence of coupling J, $\Lambda_f = \Lambda_a + \Lambda_b$, and $\omega_f = \omega_a + \omega_b$. Physically, the reorganization energy of each site corresponds to the energy the nuclei lose upon vertical excitation (i.e., excitation keeping fixed nuclear coordinates) in order to settle in the equilibrium geometry of the excited state, generally within the nanosecond timescale. Equation (2.17) shows $H_{\mathscr{S}}$ is determined not only by the energy offsets E_i of the PES but also by the reorganization energies Λ_i.

$H_{\mathscr{B}}$ is exactly given by equation (2.8*c*),

$$H_{\mathscr{B}} = \sum_{i=1}^{N_a+N_b} \frac{p_i^2}{2m_i} + \frac{m_i \omega_i^2 x_i^2}{2}. \tag{2.18}$$

That is, the bath is composed of a discrete set of harmonic oscillators of frequencies ω_i. Finally, $H_{\mathscr{S}\mathscr{B}}$ is given by,

$$H_{\mathscr{S}\mathscr{B}}(\mathbf{R}) = -\sum_{i=1}^{N_a} m_i \omega_i^2 \Delta_i x_i |a\rangle\langle a| - \sum_{i=N_a+1}^{N_a+N_b} m_i \omega_i^2 \Delta_i x_i |b\rangle\langle b|, \tag{2.19}$$

which indicates that the coupling between the system and the bath is linear in the bath coordinates and diagonal in the electronic site basis. If $\{\Delta_i\}$ are small, the dynamics of the electronic system $\mathscr{S}$ is largely described by $H_{\mathscr{S}}$, and $H_{\mathscr{S}\mathscr{B}}$ can be regarded as a perturbation. This model describes a multi-level electronic system coupled to a bath of harmonic oscillators and is a good starting point to develop theories of relaxation of open quantum systems. When a large number of bath coordinates is included, $\mathscr{B}$ can be approximated as a continuum, and the physics of the relaxation of $\mathscr{S}$ depends on the distribution of the couplings (i.e., displacements Δ_i) as a function of frequency ω_i (spectral density) [5, 6].

In QPT, we will be interested in reconstructing $\chi(T)$ for the excited electronic state dynamics of a chromophore, where the nuclear degrees of freedom of $\mathscr{B}$ are traced out (i.e., not directly measured). More specifically, we will show that certain spectroscopic signals report directly on $\chi(T)$ for the states in the SEM.

Bibliography

[1] Biggs J D and Cina J A 2009 Calculations of nonlinear wave-packet interferometry signals in the pump–probe limit as tests for vibrational control over electronic excitation transfer *J. Chem. Phys.* **131** 224302

[2] Biggs J D and Cina J A 2009 Using wave-packet interferometry to monitor the external vibrational control of electronic excitation transfer *J. Chem. Phys.* **131** 224101

[3] Cina J A, Kilin D S and Humble T S 2003 Wavepacket interferometry for short-time electronic energy transfer: multidimensional optical spectroscopy in the time domain *J. Chem. Phys.* **118** 46–61

[4] Forster T 1965 *Delocalized Excitation and Energy Transfer* vol 3 (New York: Academic Press) pp 93–137

[5] May V and Kuhn O 2004 *Charge and Energy Transfer Dynamics in Molecular Systems* (New York: Wiley-VCH)

[6] Nitzan A 2006 *Chemical Dynamics in Condensed Phases* (Oxford: Oxford University Press)

[7] Tannor D J 2007 *Introduction to Quantum Mechanics: A Time Dependent Approach* (Mill Valley, CA: University Science Books)

[8] Tully J C 1990 Molecular dynamics with electronic transitions *J. Chem. Phys.* **93** 1061–71

[9] Yarkony D R 1996 Diabolical conical intersections *Rev. Mod. Phys.* **68** 985–1013

Chapter 3

Interaction of light pulses with ensembles of chromophores: polarization gratings

In this chapter, we introduce light pulses and how they interact with an ensemble of molecules, as shown in figure 3.1. In section 3.1, we describe how light pulses produce polarization gratings. In section 3.2 we demonstrate the calculation of the induced polarization in a simple model of ideal coupled dimers. Finally, in section 3.3 we show how these polarization gratings are measured experimentally.

For concreteness, we consider a box of size L_x, L_y and L_z, respectively, which contains a homogeneous ensemble of chromophores such as those described in chapter 2, and we set the origin of our coordinate system $\mathbf{r} = (0, 0, 0)$ at the center of the box.

3.1 Laser-induced polarization gratings

We are ready to discuss the interaction of time-dependent electric fields with the ensemble of chromophores. Mathematically, we add a light-matter perturbation term to our molecular Hamiltonian of equation (2.1), so that the total Hamiltonian H reads:

$$H(\mathbf{r}, t) = H_0 + V(\mathbf{r}, t). \tag{3.1}$$

Here,

$$V(\mathbf{r}, t) = -\boldsymbol{\mu} \cdot \boldsymbol{\varepsilon}(\mathbf{r}, t) \tag{3.2}$$

is the interaction of the molecular dipole with the electric field of the light pulses. We consider N_{pulses} light pulses interacting with the sample, and we approximate them as Gaussians,

$$\boldsymbol{\varepsilon}(\mathbf{r}, t) = \sum_{n=1}^{N_{pulses}} [\varepsilon_n(t - t_n)\mathrm{e}^{\mathrm{i}\mathbf{k}_n \cdot \mathbf{r} + \mathrm{i}\phi_n}\mathbf{e}_n + \text{c.c.}], \tag{3.3}$$

$$\varepsilon_n(t) = \frac{\eta \mathrm{e}^{-t^2/2\sigma_n^2}\mathrm{e}^{-\mathrm{i}\omega_n t}}{\sqrt{2\pi\sigma_n^2}}. \tag{3.4}$$

doi:10.1088/978-0-750-31062-8ch3

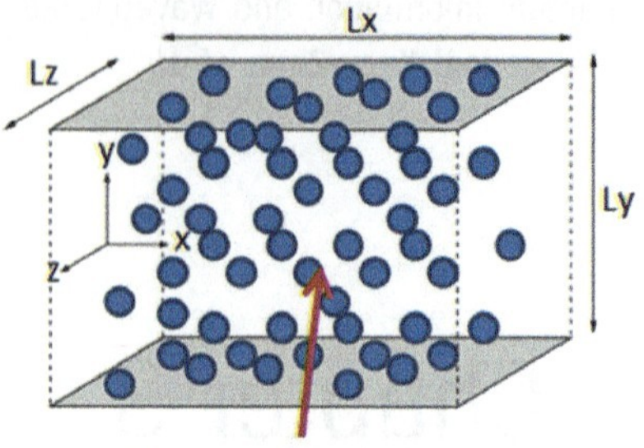

Figure 3.1. Ensemble of identical chromophores interacting with a series of, in general, noncollinear laser pulses. These interactions create a polarization grating which governs the energy exchange between the fields and the material.

Equation (3.3) describes a discrete set of pulses of weak strength η, centered about times t_n, with carrier (center) frequencies ω_n, polarization $\mathbf{e}_n$,[1] propagating along directions determined by their wavevectors $\mathbf{k}_n$. The polarization $\mathbf{e}_n$ can, in general, be a complex number to account for the possibility of circular and elliptically polarized light sources. We refer the reader to appendix A for a derivation of the pulses in equation (3.3) starting from superpositions of plane waves.

We are interested in understanding the dynamics due to the free Hamiltonian H_0. The role of $V(\mathbf{r}, t)$ is simply to induce transitions in the chromophore that will allow us to study the free evolution of its electronic degrees of freedom. Hence, η must be small enough so that $V(\mathbf{r}, t)$ acts only as a perturbation with respect to H_0, and we can use time-dependent perturbation theory. A discussion of what weak means quantitatively in our context is given in appendix B.

Due to the phases $e^{i\mathbf{k}_p\cdot\mathbf{r}+i\phi_p}$ associated with each pulse in equation (3.3), the interaction between light and molecules will have a phase that depends on the molecules' spatial locations. From perturbation theory, one can see that the pulses will induce a time-dependent polarization $\mathbf{P}$ (i.e., dipole) in each molecule, which can be expanded as [2, 4, 6],

$$\begin{aligned}\mathbf{P}(\mathbf{r}, t) &= \mathrm{Tr}\,(\boldsymbol{\mu}\rho(\mathbf{r}, t)) \\ &= \sum_s \mathbf{P}_{\mathbf{k}_s}(t)e^{i\mathbf{k}_s\cdot\mathbf{r}+i\phi_s},\end{aligned} \tag{3.5}$$

where $\boldsymbol{\mu}$ is the dipole operator acting locally at the molecule centered at $\mathbf{r}$,[2] $\rho(\mathbf{r}, t)$ is the reduced electronic density matrix of such molecule, and $\mathbf{P}_{\mathbf{k}_s}(t)$ is a

[1] Hereafter, we use the word *polarization* in two different ways: to denote (a) the orientation of oscillations of the electric field and (b) the density of electric dipole moments in a material. The meaning should be clear by the context.

[2] $\boldsymbol{\mu}$ only acts on the electronic degrees of freedom, see equation (2.9).

time-dependent polarization amplitude associated with the action of a particular combination of pulses that renders the phase $e^{i\mathbf{k}_s\cdot\mathbf{r}+i\phi_s}$. Equation (3.5) describes a spatial polarization grating in the sample. The induced *signal* wavevectors $\mathbf{k}_s$ and phases ϕ_s associated with each polarization mode (indexed by s) correspond to linear combinations of the incoming fields,

$$\mathbf{k}_s = \sum_{n=1}^{N_{pulses}} m_n \mathbf{k}_n, \tag{3.6}$$

$$\phi_s = \sum_{n=1}^{N_{pulses}} m_n \phi_n. \tag{3.7}$$

That is, each $\mathbf{k}_s$ and ϕ_s is characterized by N_{pulses} integers $m_1, m_2, \ldots, m_{N_{pulses}}$, and we consider as many modes s as linear combinations we can construct up to a given order in perturbation theory. The coefficients m_n must be integers because they indicate the excitation or de-excitation due to an integer number of photons. For instance, tracking terms *up to* first order perturbation theory, we have $m_n = 0, \pm 1$, and there are three possible values of s, corresponding to a zeroth order and two first order contributions. Remarkably, the polarization grating described by equation (3.5) encodes information about the wavevector modes of the incoming fields that created it. As we will show next, this spatial grating determines how energy is exchanged between the sample and the different pulses, and hence, dictates the nature of certain spectroscopic signals that depend directly on the Fourier components $\mathbf{P}_{\mathbf{k}_s}(t)$. The calculation of these components is the central problem of nonlinear optical spectroscopy and will be thoroughly illustrated with the next section.

3.2 Induced linear and nonlinear polarization in an ideal coupled dimer

In this section, we explicitly calculate the polarization induced by short laser pulses in an ensemble of chromophores. This will give us a better feel for the derivations in the following chapters. However, the reader might skip to section 3.3 and only use it as a reference when needed.

The model we consider is a simplification of the coupled dimer. Namely, we investigate an ensemble of vibrationless coupled dimers where the electronic system $\mathcal{S}$ does not interact with the vibrations $\mathcal{B}$ (i.e., $\mathcal{S}$ is an *isolated* system). This means that equations (1.1) and (2.17)–(2.19) hold, but we assume that $H_{\mathcal{S}\mathcal{B}} = 0$ by setting $\{\Delta_i\} = 0$. Physically, this situation corresponds to dimers made up of rigid monomers whose excited states do not differ in geometry compared to their ground states. Assume that at the beginning, at $t = 0$, the electronic state of each molecule is $|\Psi(\mathbf{r}, 0)\rangle \equiv |\Psi_0\rangle = |g\rangle$ for all $\mathbf{r}$ values corresponding to the centers of mass of the molecules. Let the array interact with two pulses, which we label P and P' (we will identify them later as the pump and the probe pulses, respectively), with electric field,

$$\boldsymbol{\varepsilon}(\mathbf{r}, t) = \sum_{n=P,P'} [\varepsilon_n(t - t_n)e^{i\mathbf{k}_n\cdot\mathbf{r}+i\phi_n}\mathbf{e}_n + \text{c.c.}]. \tag{3.8}$$

The propagation directions of the pulses are different, $\mathbf{k}_P \neq \mathbf{k}_{P'}$. This scenario is depicted in figure 4.4 in chapter 4, where we consider the general theory of PP' spectroscopy.

Our goal (see section 3.3) is to calculate the polarization component along the $\mathbf{k}_{P'}$ direction, that is, $\mathbf{P}_{\mathbf{k}_{P'}}(t)$, up to $O(\eta^3)$. For that, we consider the following assumptions,

$$\text{Assumption 1: } t_P - \sigma_P \gg 0, \tag{3.9}$$

$$\text{Assumption 2: } T \equiv t_{P'} - t_P \gg \{\sigma_P, \sigma_{P'}\}, \tag{3.10}$$

$$\text{Assumption 3: } t \gg t_{P'} + \sigma_{P'}. \tag{3.11}$$

Assumption 1 is not really necessary; it simply allows us to regard $t = 0$ as the initial time in the absence of perturbations, but clearly, the time origin is arbitrary. We keep this assumption throughout the book unless indicated otherwise, the purpose being to ease the notation and easily keep track of phases, although a careful book-keeping of the latter should render the same physical observables. Assumption 2 guarantees that the two pulses do not overlap, and that P arrives before P'. T is called the *waiting time*. Finally, assumption 3 asks us to think about the polarization after both pulses have acted on the system, ignoring transient effects. Such an assumption will be dropped in other examples in this book, as it is precisely the transients in the polarization that contribute to various spectroscopic signals once they have interfered with another reference pulse. For this introductory calculation, however, it will clear our minds to keep the last assumption.

3.2.1 Eigenvalues, eigenvectors and energy scales

We begin by deriving expressions for the eigenvalues and eigenvectors of $H_{\mathcal{S}}$ (see equations (2.8*b*) and (2.2)). The Hamiltonian $H_{\mathcal{S}}$, represented in the $\{|g\rangle, |a\rangle, |b\rangle, |f\rangle\}$ basis, is given by,

$$H_{\mathcal{S}} = \begin{pmatrix} \omega_g & 0 & 0 & 0 \\ 0 & \omega_a & J & 0 \\ 0 & J & \omega_b & 0 \\ 0 & 0 & 0 & \omega_f \end{pmatrix}, \tag{3.12}$$

which upon diagonalization yields the eigenvectors $\{|g\rangle, |\alpha\rangle, |\beta\rangle, |f\rangle\}$ and eigenvalues $\{\omega_g, \omega_\alpha, \omega_\beta, \omega_f\}$. Clearly, $|g\rangle$ and $|f\rangle$ remain eigenstates of $H_{\mathcal{S}}$ since they do not couple to other states. The diagonalization of the SEM is just the standard two-level system problem with real parameters, whose solutions can be expressed in terms of the angle $\theta = \frac{1}{2}\arctan(\frac{J}{\delta})$, where $\delta = \omega_a - \omega_b$. Then, the eigenvectors are $|\alpha\rangle = -\sin\theta|a\rangle + \cos\theta|b\rangle$ and $|\beta\rangle = \cos\theta|a\rangle + \sin\theta|b\rangle$, whereas the eigenvalues are given by $\omega_\alpha = \overline{\omega} - \delta\sec 2\theta$ and $\omega_\beta = \overline{\omega} + \delta\sec 2\theta$, where $\overline{\omega} = \frac{1}{2}(\omega_a + \omega_b)$ [2, 3]. Notice that $\omega_f = \omega_a + \omega_b = \omega_\alpha + \omega_\beta$, which is again a statement of the lack of interaction between the two excitations in this doubly excited state. The dipoles in equation (2.9) transform as,

$$\begin{bmatrix} \boldsymbol{\mu}_{\alpha g} \\ \boldsymbol{\mu}_{\beta g} \end{bmatrix} = \begin{bmatrix} -\sin\theta & \cos\theta \\ \cos\theta & \sin\theta \end{bmatrix} \begin{bmatrix} \boldsymbol{\mu}_{ag} \\ \boldsymbol{\mu}_{bg} \end{bmatrix}, \tag{3.13}$$

$$\begin{bmatrix} \boldsymbol{\mu}_{f\alpha} \\ \boldsymbol{\mu}_{f\beta} \end{bmatrix} = \begin{bmatrix} \cos\theta & -\sin\theta \\ \sin\theta & \cos\theta \end{bmatrix} \begin{bmatrix} \boldsymbol{\mu}_{ag} \\ \boldsymbol{\mu}_{bg} \end{bmatrix}. \tag{3.14}$$

To get a rough idea of the energy scales, we also compute numerical estimates using the values, $\omega_g = 0$, $\omega_a = 12\,719$ cm^{-1}, $\omega_b = 12\,881$ cm^{-1}, $J = 120$ cm^{-1}. Similar energy scales are reported for photosynthetic complexes [5, 9]. These values yield $\theta = -0.319$, $\omega_g = 0$, $\omega_\alpha = 12\,655$ cm^{-1}, $\omega_\beta = 12\,945$ cm^{-1}, $\omega_f = 25\,600$ cm^{-1}, $|\alpha\rangle = -0.88|a\rangle + 0.47|b\rangle$, and $|\beta\rangle = 0.47|a\rangle + 0.88|b\rangle$.

3.2.2 Induced polarization

For any system in the Condon approximation, $\boldsymbol{\mu}$ acts only on electronic degrees of freedom. Therefore, keeping track of the (reduced) electronic state of the molecules in the ensemble suffices to compute $\mathbf{P}_{\mathbf{k}_{P'}}(t)$. In this example, the fact that $H_{\mathscr{S}\mathscr{B}} = 0$ simplifies the problem because the system $\mathscr{S}$ evolves independently from the bath $\mathscr{B}$. If $|\Psi\rangle$ denotes a state of $\mathscr{S}$, its evolution via H_0 only depends on $H_{\mathscr{S}}$, $e^{-iH_0t}|\Psi\rangle = e^{-iH_{\mathscr{S}}t}|\Psi\rangle$. Even though in general we will write $|\Psi\rangle$ to denote a wavefunction in the Hilbert space of $\mathscr{S}$ and $\mathscr{B}$, for this particular problem, the system $\mathscr{S}$ is isolated, so calculating the wavefunction for $\mathscr{S}$ suffices. Let $|\Psi(\mathbf{r}, t)\rangle$ be the electronic state of a molecule at position $\mathbf{r}$ after the two pulses of light (assumption 3, equation (3.11)). Using time-dependent perturbation theory, we may expand $|\Psi(\mathbf{r}, t)\rangle$ to the following expression, which we will simplify in the next chapters:

$$\begin{aligned}
|\Psi(\mathbf{r},t)\rangle = |\Psi_0(t)\rangle &+ \sum_{n=P,P'} e^{i(\mathbf{k}_n\cdot\mathbf{r}+\phi_n)}|\Psi_{+n}(t)\rangle + \sum_{n=P,P'} e^{-i(\mathbf{k}_n\cdot\mathbf{r}+\phi_n)}|\Psi_{-n}(t)\rangle \\
&+ \sum_{n,n'=P,P'} \Big[e^{i(\mathbf{k}_n\cdot\mathbf{r}+\phi_n)+i(\mathbf{k}_{n'}\cdot\mathbf{r}+\phi_{n'})}|\Psi_{+n+n'}(t)\rangle \\
&+ e^{i(\mathbf{k}_n\cdot\mathbf{r}+\phi_n)-i(\mathbf{k}_{n'}\cdot\mathbf{r}+\phi_{n'})}|\Psi_{+n-n'}(t)\rangle + e^{-i(\mathbf{k}_n\cdot\mathbf{r}+\phi_n)+i(\mathbf{k}_{n'}\cdot\mathbf{r}+\phi_{n'})}|\Psi_{-n+n'}(t)\rangle \\
&+ e^{-i(\mathbf{k}_n\cdot\mathbf{r}+\phi_n)-i(\mathbf{k}_{n'}\cdot\mathbf{r}+\phi_{n'})}|\Psi_{-n-n'}(t)\rangle \Big] \\
&+ \sum_{n,n',n''=P,P'} \Big[e^{i(\mathbf{k}_n\cdot\mathbf{r}+\phi_n)+i(\mathbf{k}_{n'}\cdot\mathbf{r}+\phi_{n'})+i(\mathbf{k}_{n''}\cdot\mathbf{r}+\phi_{n''})}|\Psi_{+n+n'+n''}(t)\rangle \\
&+ e^{-i(\mathbf{k}_n\cdot\mathbf{r}+\phi_n)+i(\mathbf{k}_{n'}\cdot\mathbf{r}+\phi_{n'})+i(\mathbf{k}_{n''}\cdot\mathbf{r}+\phi_{n''})}|\Psi_{-n+n'+n''}(t)\rangle \\
&+ e^{i(\mathbf{k}_n\cdot\mathbf{r}+\phi_n)-i(\mathbf{k}_{n'}\cdot\mathbf{r}+\phi_{n'})+i(\mathbf{k}_{n''}\cdot\mathbf{r}+\phi_{n''})}|\Psi_{+n-n'+n''}(t)\rangle
\end{aligned}$$

$$+ e^{i(\mathbf{k}_n\cdot\mathbf{r}+\phi_n)+i(\mathbf{k}_{n'}\cdot\mathbf{r}+\phi_{n'})-i(\mathbf{k}_{n''}\cdot\mathbf{r}+\phi_{n''})}|\Psi_{+n+n'-n''}(t)\rangle$$

$$+ e^{-i(\mathbf{k}_n\cdot\mathbf{r}+\phi_n)-i(\mathbf{k}_{n'}\cdot\mathbf{r}+\phi_{n'})+i(\mathbf{k}_{n''}\cdot\mathbf{r}+\phi_{n''})}|\Psi_{-n-n'+n''}(t)\rangle$$

$$+ e^{-i(\mathbf{k}_n\cdot\mathbf{r}+\phi_n)+i(\mathbf{k}_{n'}\cdot\mathbf{r}+\phi_{n'})-i(\mathbf{k}_{n''}\cdot\mathbf{r}+\phi_{n''})}|\Psi_{-n+n'-n''}(t)\rangle$$

$$+ e^{i(\mathbf{k}_n\cdot\mathbf{r}+\phi_n)-i(\mathbf{k}_{n'}\cdot\mathbf{r}+\phi_{n'})-i(\mathbf{k}_{n''}\cdot\mathbf{r}+\phi_{n''})}|\Psi_{+n-n'-n''}(t)\rangle$$

$$+ e^{-i(\mathbf{k}_n\cdot\mathbf{r}+\phi_n)-i(\mathbf{k}_{n'}\cdot\mathbf{r}+\phi_{n'})-i(\mathbf{k}_{n''}\cdot\mathbf{r}+\phi_{n''})}|\Psi_{-n-n'-n''}(t)\rangle\Big] + O(\eta^4). \quad (3.15)$$

The notation is such that, for instance, $|\Psi_{+P+P'-P'}(t)\rangle$ denotes the wavefunction for the molecule at $\mathbf{r} = 0$ resulting from the interaction with ε_P first, then $\varepsilon_{P'}$ and finally $\varepsilon^*_{P'}$, whereas $|\Psi_{+P-P'+P'}(t)\rangle$ represents the perturbations in the order ε_P, $\varepsilon^*_{P'}$ and $\varepsilon_{P'}$. The factor,

$$e^{i(\mathbf{k}_n\cdot\mathbf{r}+\phi_n)+i(\mathbf{k}_{n'}\cdot\mathbf{r}+\phi_{n'})-i(\mathbf{k}_{n''}\cdot\mathbf{r}+\phi_{n''})}, \quad (3.16)$$

accounts for the spatial dependence of a given term in the wavefunction (and does not depend on the time-ordering of the interactions). These are relative phases between different terms in the wavefunction, so they *do* matter. We recognize that $\mathbf{k}_n$ always comes together with $\varepsilon_n \propto e^{-i\omega_n(t-t_n)}$, whereas $-\mathbf{k}_n$ with $\varepsilon^*_n \propto e^{i\omega_n(t-t_n)}$ (equations (3.3) and (3.4)).

From equation (3.5), the induced polarization takes the form,

$$\mathbf{P}(\mathbf{r}, t) = \langle\Psi(\mathbf{r}, t)|\boldsymbol{\mu}|\Psi(\mathbf{r}, t)\rangle, \quad (3.17)$$

of which we are only interested in the component $\mathbf{P}_{\mathbf{k}_{P'}}(t)$. Let us develop each of the perturbation terms of $|\Psi(\mathbf{r}, t)\rangle$ that are relevant for the calculation of this component.

Zeroth order contribution:

$$|\Psi_0(t)\rangle \equiv e^{-iH_0 t}|\Psi_0\rangle \quad (3.18a)$$

$$= e^{-i\omega_g t}|g\rangle, \quad (3.18b)$$

where we have used $H_0|\Psi_0\rangle = \omega_g|g\rangle$.

First order contributions:

There are four first order perturbation terms, $|\Psi_{\pm P}(t)\rangle$ and $|\Psi_{\pm P'}(t)\rangle$. We start by calculating $|\Psi_{+n}(t)\rangle$ in general, for $n = P, P'$,[3]

[3] Note that the perturbative wavefunctions in this book are not normalized.

$$|\Psi_{+n}(t)\rangle \equiv -\mathrm{i}\int_0^t \mathrm{d}t' \mathrm{e}^{-\mathrm{i}H_0(t-t')}\{-\boldsymbol{\mu}\cdot\mathbf{e}_n\varepsilon_n(t'-t_n)\}\mathrm{e}^{-\mathrm{i}H_0t'}|\Psi_0\rangle \tag{3.19a}$$

$$\approx \mathrm{i}\sum_{q=\alpha,\beta}|q\rangle\int_{-\infty}^{\infty}\mathrm{d}t'\mathrm{e}^{-\mathrm{i}\omega_q(t-t')}\{\boldsymbol{\mu}_{qg}\cdot\mathbf{e}_n\varepsilon_n(t'-t_n)\}\mathrm{e}^{-\mathrm{i}\omega_g t'} \tag{3.19b}$$

$$= \mathrm{i}\sum_{q=\alpha,\beta}\underbrace{|q\rangle}_{\text{final state}}\left(\underbrace{\mathrm{e}^{-\mathrm{i}\omega_q(t-t_n)}}_{\text{evolution via } H_0 \text{ in } |q\rangle}\right)$$

$$\times\left(\underbrace{\Omega^n_{qg}}_{\text{transition amplitude from } |g\rangle \text{ to } |q\rangle \text{ via } \varepsilon_n}\right)\left(\underbrace{\mathrm{e}^{-\mathrm{i}\omega_g t_n}}_{\text{evolution via } H_0 \text{ in } |g\rangle}\right). \tag{3.19c}$$

From equations (3.19*a*) to (3.19*b*), we have extended the integral to $t' \in (-\infty, \infty)$ using assumptions 1 and 3 (equations (3.9) and (3.11)). We have also introduced Ω^n_{ij}, the transition probability amplitude to go from state $|j\rangle$ to state $|i\rangle$ via ε_n,

$$\Omega^n_{ij} = \tilde{\varepsilon}_n(\omega_{ij})\boldsymbol{\mu}_{ij}\cdot\mathbf{e}_n, \tag{3.20}$$

which depends on the transition dipole projected onto the pulse polarization $\boldsymbol{\mu}_{ij}\cdot\mathbf{e}_n$ and the amplitude of the electric field $\tilde{\varepsilon}_n(\omega_{ij})$ at the given transition energy. The latter is a Fourier transform of the Gaussian pulse,

$$\tilde{\varepsilon}_n(\omega) = \int_{-\infty}^{\infty}\mathrm{d}t\,\mathrm{e}^{\mathrm{i}\omega t}\varepsilon_n(t) = \eta\mathrm{e}^{-(\omega-\omega_n)^2\sigma_n^2/2}. \tag{3.21}$$

Note that the carrier frequency ω_n of the pulses denotes the center frequency of the distribution $\tilde{\varepsilon}(\omega)$, whereas the pulse-width in time σ_n determines the width in frequency σ_n^{-1}. It will also be convenient to define the transition amplitudes due to the conjugate term ε_n^*,[4]

$$\begin{aligned}\Omega^{\bar{n}}_{ji} &= \tilde{\varepsilon}^*_n(\omega_{ij})\boldsymbol{\mu}_{ji}\cdot\mathbf{e}^*_n\\ &= (\Omega^n_{ij})^*.\end{aligned} \tag{3.22}$$

Equation (3.19*c*) conveys a simple physical picture consisting of two free evolutions with respect to H_0 (or $H_{\mathscr{S}}$, in this case), connected by a transition due to the field. The transition amplitude Ω^n_{qg} depends both on the alignment of the dipole moment of each transition $\boldsymbol{\mu}_{qg}$ with the polarization of the field $\mathbf{e}_n$ as well as on the amplitude of the pulse at the transition energy ω_{qg}. Importantly, $|\Psi_{+n}(t)\rangle$ is a coherent superposition of $|\alpha\rangle$ and $|\beta\rangle$ as long as $\tilde{\varepsilon}_n(\omega_{\alpha g})$ and $\tilde{\varepsilon}_n(\omega_{\beta g})$ are both sufficiently large (that is, the pulse sufficiently broadband). The free evolutions go from 0 to t_n and from t_n to t, where t_n is the center time of the pulse n. This allows for the

[4] Here, $\tilde{\varepsilon}_n(\omega) = \tilde{\varepsilon}^*_n(\omega)$ for equation (3.4). For more general pulse forms $\varepsilon_n(t-t_n)$, however, this need not be the case. Hence, we shall keep the distinction to remind ourselves that $\tilde{\varepsilon}_n(\omega)$ arises from the action of $\varepsilon_n(t-t_n)$, whereas $\tilde{\varepsilon}^*_n(\omega)$ comes from $\varepsilon^*_n(t-t_n)$.

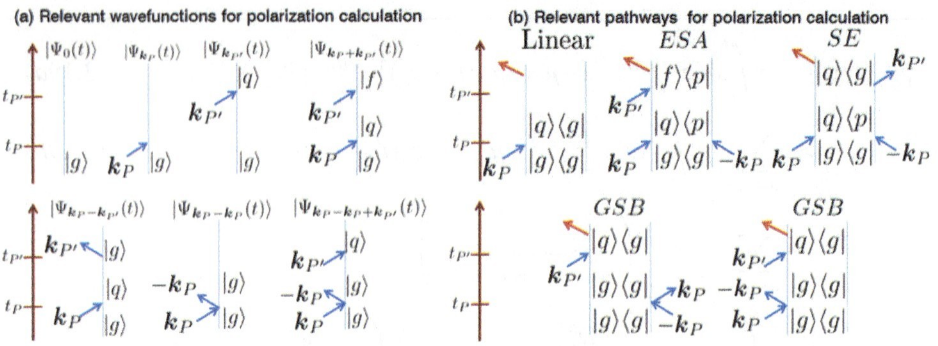

Figure 3.2. (a) Feynman diagrams for the evolution of wavefunctions associated with different actions of pulses. (b) Double-Sided-Feynman-Diagrams (DS-FDs) for the evolution of different coherences and populations associated with the pulses. Red arrows indicate measurement of polarization.

loose interpretation that the pulse acts impulsively at t_n even when it is not necessarily a δ-pulse. A moment of reflection cautions on the limits of this observation: it only holds if $t \gg t_n + \sigma_n$, which clearly holds if t satisfies Assumption 3 (equation (3.11)). Otherwise, for times $t \in [t_n - 3\sigma_n, t_n + 3\sigma_n]$ the integral in equation (3.19*a*) cannot be taken to ∞ and transient effects may be important. These restrictions are a consequence of the time–energy uncertainty principle. However, when equation (3.19*a*) is valid, the particular motif of,

$$\text{(FREE EVOLUTION)-(INTERACTION WITH PULSE)-(FREE EVOLUTION)}, \tag{3.23}$$

is one of the most important intuitions for the design of ultrafast spectroscopy measurements, and as we shall see, will be a recurring structure for general multipulse schemes. It states that the quantum state of the system in question can be understood as a free evolution due to H_0 interrupted by impulsive transitions induced by the pulses, confirming our intuition from chapter 2. This intuition is schematically depicted in figure 3.2(*a*) as Feynman diagrams for a series of representative wavefunctions that are involved in the calculation of a desired polarization component (figure 3.2(*b*) will make sense by the end of this subsection).

We now proceed to compute $|\Psi_{-n}(t)\rangle$,

$$\begin{aligned}|\Psi_{-n}(t)\rangle &= -\mathrm{i}\int_0^t \mathrm{d}t' \mathrm{e}^{-\mathrm{i}H_0(t-t')}\{-\boldsymbol{\mu}\cdot \mathbf{e}_n^* \varepsilon_n^*(t'-t_n)\}\mathrm{e}^{-\mathrm{i}H_0 t'}|\Psi_0\rangle \\ &= \mathrm{i}\sum_{q=\alpha,\beta}|q\rangle \mathrm{e}^{-\mathrm{i}\omega_q(t-t_n)}\Omega_{qg}^{\bar{n}}\mathrm{e}^{-\mathrm{i}\omega_g t_n} \\ &\approx 0,\end{aligned} \tag{3.24}$$

where we have noticed that $\Omega_{qg}^{\bar{n}} \propto \tilde{\varepsilon}_n^*(-\omega_{qg}) \approx 0$, because $\tilde{\varepsilon}_n(\pm\omega_{qg}) = \tilde{\varepsilon}_n^*(\pm\omega_{qg}) = \eta \mathrm{e}^{-(\omega_n \mp \omega_{qg})^2\sigma_n^2/2}$, which implies that $\tilde{\varepsilon}_n^*(-\omega_{qg}) \ll \tilde{\varepsilon}_n^*(\omega_{qg})$. This observation, which holds whenever a resonant process dominates with respect to an off-resonant one, justifies the following very important rotating-wave approximation (RWA) [6],

$$\begin{aligned}
&\varepsilon_n(t-t_n) \propto \mathrm{e}^{-\mathrm{i}\omega_n(t-t_n)} \quad \begin{cases} \text{excites ket amplitude to higher energy states} \\ \text{de-excites bra amplitude to lower energy states} \end{cases} \\
&\varepsilon_n^*(t-t_n) \propto \mathrm{e}^{\mathrm{i}\omega_n(t-t_n)} \quad \begin{cases} \text{de-excites ket amplitude to lower energy states} \\ \text{excites bra amplitude to higher energy states} \end{cases}
\end{aligned} \tag{3.25}$$

For later discussions, we have also mentioned the rule for perturbations in the bra space, which being the dual of the ket space, must follow the complex conjugate of the corresponding rules. When neither ε_n nor ε_n^* can induce a particular transition resonantly, the RWA might not hold, and it might be necessary to include both terms. This is, however, not the case in this book, where we deal with resonant interactions, and we simply ignore the so-called *counterrotating terms*,

$$\begin{aligned}
&\text{Only consider resonant rotating terms } \Omega_{ij}^{n},\ \Omega_{ji}^{\bar{n}} \text{ for } \omega_{ij} > 0, \\
&\text{Ignore counterrotating terms } \Omega_{ij}^{n},\ \Omega_{ji}^{\bar{n}} \text{ for } \omega_{ij} < 0.
\end{aligned} \tag{3.26}$$

By applying the RWA, we may immediately discard several terms in the expansion of equation (3.15) as negligible,

$$|\Psi_{-n\pm n'}(t)\rangle,\ |\Psi_{-n\pm n'\pm n''}(t)\rangle \approx 0, \tag{3.27}$$

since the first perturbation associated with ε_n^* can only de-excite a ket, but this is impossible if the state at the beginning is in $|g\rangle$.

Second order contributions:

Let us now turn to some second order wavefunctions. We start with,

$$\begin{aligned}
|\Psi_{+P-P'}(t)\rangle \equiv (-\mathrm{i})^2 \int_0^t \mathrm{d}t' \int_0^{t'} \mathrm{d}t'' \mathrm{e}^{-\mathrm{i}H_0(t-t')}\{-\boldsymbol{\mu}\cdot\mathbf{e}_{P'}^*\varepsilon_{P'}^*(t'-t_{P'})\} \\
\times\, \mathrm{e}^{-\mathrm{i}H_0(t'-t'')}\{-\boldsymbol{\mu}\cdot\mathbf{e}_P\varepsilon_P(t''-t_P)\}\mathrm{e}^{-\mathrm{i}H_0t''}|\Psi_0\rangle
\end{aligned} \tag{3.28a}$$

$$\begin{aligned}
&\approx (-\mathrm{i})^2 \int_{-\infty}^{\infty} \mathrm{d}t' \mathrm{e}^{-\mathrm{i}H_0(t-t')}\{-\boldsymbol{\mu}\cdot\mathbf{e}_{P'}^*\varepsilon_{P'}^*(t'-t_{P'})\}|\Psi_{+P}(t')\rangle \\
&= -|g\rangle \sum_{q=\alpha,\beta} \mathrm{e}^{-\mathrm{i}\omega_g(t-t_{P'})}\Omega_{gq}^{\overline{P}'}\mathrm{e}^{-\mathrm{i}\omega_q T}\Omega_{qg}^{P}\mathrm{e}^{-\mathrm{i}\omega_g t_P}.
\end{aligned} \tag{3.28b}$$

Here, we have extended both integrals to be from $-\infty$ to ∞ owing to the three assumptions, equations (3.9)–(3.11). This allows approximating the nested integrals from a second order perturbation theory as two sequential first order perturbation calculations like the one from equation (3.19*a*). The interpretation is analogous to the first order calculation. Equation (3.28*a*) denotes a free evolution from 0 to t_P in $|g\rangle$, an excitation via ε_P from $|g\rangle$ to $|q\rangle$, an evolution from t_P to $t_{P'}$ in $|q\rangle$, a de-excitation via $\varepsilon^*_{P'}$ back to $|g\rangle$, and a final evolution from $t_{P'}$ to t in $|g\rangle$. Crucially, amplitude information on the coherent dynamics that occurs in $|\alpha\rangle$ and $|\beta\rangle$ during the waiting time T, namely, the phase information $\mathrm{e}^{-\mathrm{i}\omega_q T}$, is transferred into the final amplitude in $|g\rangle$.

Very similarly,

$$
\begin{aligned}
|\Psi_{+P+P'}(t)\rangle &= (-\mathrm{i})^2 \int_0^t \mathrm{d}t' \int_0^{t'} \mathrm{d}t'' \mathrm{e}^{-\mathrm{i}H_0(t-t')}\{-\boldsymbol{\mu}\cdot\mathbf{e}_{P'}\varepsilon_{P'}(t'-t_{P'})\} \\
&\quad \times \mathrm{e}^{-\mathrm{i}H_0(t'-t'')}\{-\boldsymbol{\mu}\cdot\mathbf{e}_P\varepsilon_P(t''-t_P)\}\mathrm{e}^{-\mathrm{i}H_0 t''}|\Psi_0\rangle \qquad (3.29a)\\
&\approx -|f\rangle \sum_{q=\alpha,\beta} \mathrm{e}^{-\mathrm{i}\omega_f(t-t_{P'})}\Omega^{P'}_{fq}\mathrm{e}^{-\mathrm{i}\omega_q(t_{P'}-t_P)}\Omega^{P}_{qg}\mathrm{e}^{-\mathrm{i}\omega_g t_P}, \qquad (3.29b)
\end{aligned}
$$

where instead of de-exciting amplitude with $\varepsilon^*_{P'}$ in the last step, amplitude is excited to $|f\rangle$ via $\varepsilon_{P'}$.

Finally,

$$
\begin{aligned}
|\Psi_{+n-n}(t)\rangle &\equiv (-\mathrm{i})^2 \int_0^t \mathrm{d}t' \int_0^{t'} \mathrm{d}t'' \mathrm{e}^{-\mathrm{i}H_0(t-t')}\{-\boldsymbol{\mu}\cdot\mathbf{e}^*_n\varepsilon^*_n(t'-t_n)\} \\
&\quad \times \mathrm{e}^{-\mathrm{i}H_0(t'-t'')}\{-\boldsymbol{\mu}\cdot\mathbf{e}_n\varepsilon_n(t''-t_n)\}\mathrm{e}^{-\mathrm{i}H_0 t''}|\Psi_0\rangle \qquad (3.30a)\\
&\approx -|g\rangle \mathrm{e}^{-\mathrm{i}\omega_g(t-t_n)} \sum_{q=\alpha,\beta} \frac{\Omega^{\bar{n}}_{gq}\Omega^{n}_{qg}}{2}[1-\mathrm{erf}(\mathrm{i}(\omega_{qg}-\omega_n)\sigma_n)]\mathrm{e}^{-\mathrm{i}\omega_g t_n}, \qquad (3.30b)
\end{aligned}
$$

for $n = P, P'$. Here, the nested integrals cannot be trivially decoupled as in equations (3.28*a*) or (3.29*a*) because they involve a second order perturbation with the same pulse n. However, assumptions 1 and 3 (equations (3.9) and (3.11)) imply that $\int_0^t \mathrm{d}t' \int_0^{t'} \mathrm{d}t'' \approx \int_{-\infty}^{\infty} \mathrm{d}t' \int_{-\infty}^{t'} \mathrm{d}t''$. Notice the $\frac{1}{2}$ factor as well as the erf (error function) terms. These are just a consequence of the fact that the pulse can only de-excite amplitude *after* it has created amplitude in the excited states[5].

[5] In order to evaluate nested integrals like equation (3.30*a*), change variables $s = t' - t''$, so that $\int_{-\infty}^{\infty} \mathrm{d}t' \int_{-\infty}^{t'} \mathrm{d}t'' = \int_0^{\infty} \mathrm{d}s \int_{-\infty}^{\infty} \mathrm{d}t''$. More explicitly, the following result may become handy throughout the book (for the derivation, see for instance, the online supporting information for [1]),

$$
\begin{aligned}
I(\alpha,\beta) &= \int_{-\infty}^{\infty} \mathrm{d}\tau_2 \int_{-\infty}^{\tau_2} \mathrm{d}\tau_1 \exp\left\{-\frac{\tau_1^2}{2\sigma^2} + \mathrm{i}\alpha\tau_1 - \frac{\tau_2^2}{2\sigma^2} - \mathrm{i}\beta\tau_2\right\} \\
&= \pi\sigma^2 \exp\left\{-\frac{(\alpha^2+\beta^2)\sigma^2}{2}\right\}\left(1-\mathrm{erf}\left(\frac{\mathrm{i}(\alpha+\beta)\sigma}{2}\right)\right).
\end{aligned}
$$

Note that even if $\mathrm{erf}(\frac{\mathrm{i}(\alpha+\beta)\sigma}{2})$ increases rapidly as α, β, and σ increase, $\exp\{-\frac{(\alpha^2+\beta^2)\sigma^2}{2}\}$ decreases even faster, bounding the result at $|I(\alpha,\beta)| \leqslant \pi\sigma^2$.

Third order contributions:
Similarly, owing to the three assumptions (equations (3.9)–(3.11)),

$$|\Psi_{+P-P+P'}(t)\rangle \equiv (-\mathrm{i})\int_0^t \mathrm{d}t' \mathrm{e}^{-\mathrm{i}H_0(t-t')}\{-\boldsymbol{\mu}\cdot\mathbf{e}_{P'}\varepsilon_{P'}(t'-t_{P'})\}|\Psi_{P-P}(t')\rangle \quad (3.31a)$$

$$\approx -\mathrm{i}\sum_{q=\alpha,\beta}|q\rangle \mathrm{e}^{-\mathrm{i}\omega_q(t-t_{P'})}\Omega^{P'}_{qg}\mathrm{e}^{-\mathrm{i}\omega_g(t_{P'}-t_P)}$$

$$\times \sum_{r=\alpha,\beta}\frac{\Omega^{\overline{P}}_{gr}\Omega^{P}_{rg}}{2}[1-\mathrm{erf}(\mathrm{i}(\omega_{qg}-\omega_P)\sigma_P)]\mathrm{e}^{-\mathrm{i}\omega_g t_P}, \quad (3.31b)$$

where the ground-state amplitude from $|\Psi_{+P-P}(t')\rangle$ gets excited to the SEM states $|q\rangle$ via $\varepsilon_{P'}$. Notice that each perturbation due to the light contributes a factor of $+\mathrm{i}$.

Several other wavefunctions up to $O(\eta^3)$ can be calculated, but it turns out that in the RWA, equations (3.18*a*)–(3.19*a*), (3.24), (3.28*a*)–(3.31*a*) suffice to calculate $\mathbf{P}_{\mathbf{k}_{P'}}(t)$ to $O(\eta^3)$. They have been schematically depicted as Feynman diagrams in figure 3.2(*a*). Note also that we have not calculated several wavefunctions such as $|\Psi_{+P'+P}(t)\rangle$, where P' acts before P, which can be set to 0 due to assumption 2 (equation (3.10)). Expanding $\mathbf{P}_{\mathbf{k}_{P'}}(t)$ in powers of η,

$$\mathbf{P}_{\mathbf{k}_{P'}}(t) = \mathbf{P}^{(1)}_{\mathbf{k}_{P'}}(t) + \mathbf{P}^{(3)}_{\mathbf{k}_{P'}}(t) + O(\eta^5),$$

where we have noticed that the polarization only has components that are odd powers of η because the dipole operator from (2.9) only couples manifolds that differ by one excitation. Collecting our previous results, the linear polarization at $\mathbf{k}_{P'}$, which is associated with the linear absorption of pulse P' by the sample, from equations (3.17), (3.15) and (3.19), is given by

$$\mathbf{P}^{(1)}_{\mathbf{k}_{P'}}(t) = \langle\Psi_0(t)|\boldsymbol{\mu}|\Psi_{P'}(t)\rangle$$

$$\approx \mathrm{i}\sum_{q=\alpha,\beta}\boldsymbol{\mu}_{qg}\Omega^{P'}_{qg}\mathrm{e}^{-\mathrm{i}\omega_{qg}t}, \quad (3.32)$$

whereas the third order polarization, related to the changes in absorption of P' due to the previous action of P on the sample, is,

$$\mathbf{P}^{(3)}_{\mathbf{k}_{P'}}(t) = \langle\Psi_{+P}(t)|\boldsymbol{\mu}|\Psi_{+P+P'}(t)\rangle + \langle\Psi_{+P-P'}(t)|\boldsymbol{\mu}|\Psi_{+P}(t)\rangle$$
$$+ \langle\Psi_{+P-P}(t)|\boldsymbol{\mu}|\Psi_{+P'}(t)\rangle + \langle\Psi_0(t)|\boldsymbol{\mu}|\Psi_{+P-P+P'}(t)\rangle. \quad (3.33)$$

The components in equation (3.33) are given by

$$\langle\Psi_{+P}(t)|\boldsymbol{\mu}|\Psi_{+P+P'}(t)\rangle = \mathrm{i}\sum_{p,q=\alpha,\beta}\boldsymbol{\mu}_{pf}\Omega^{P'}_{fq}\Omega^{P}_{qg}\Omega^{\overline{P}}_{gp}\mathrm{e}^{-\mathrm{i}\omega_{fp}\bar{t}}\mathrm{e}^{-\mathrm{i}\omega_{qp}T}, \tag{3.34a}$$

$$\langle\Psi_{+P-P'}(t)|\boldsymbol{\mu}|\Psi_{+P}(t)\rangle = -\mathrm{i}\sum_{p,q=\alpha,\beta}\boldsymbol{\mu}_{gq}\Omega^{P}_{qg}\Omega^{\overline{P}}_{gp}\Omega^{P'}_{pg}\mathrm{e}^{-\mathrm{i}\omega_{qg}\bar{t}}\mathrm{e}^{-\mathrm{i}\omega_{qp}T}, \tag{3.34b}$$

$$\langle\Psi_{+P-P}(t)|\boldsymbol{\mu}|\Psi_{+P'}(t)\rangle = -\frac{\mathrm{i}}{2}\sum_{p,q=\alpha,\beta}\boldsymbol{\mu}_{gq}\Omega^{P'}_{qg}\Omega^{\overline{P}}_{gp}\Omega^{P}_{pg}[1+\mathrm{erf}(\mathrm{i}(\omega_{qg} - \omega_P)\sigma_P)]\mathrm{e}^{-\mathrm{i}\omega_{qg}\bar{t}}, \tag{3.34c}$$

$$\langle\Psi_{0}(t)|\boldsymbol{\mu}|\Psi_{+P-P+P'}(t)\rangle = -\frac{\mathrm{i}}{2}\sum_{p,q=\alpha,\beta}\boldsymbol{\mu}_{gq}\Omega^{P'}_{qg}\Omega^{\overline{P}}_{gp}\Omega^{P}_{pg}[1-\mathrm{erf}(\mathrm{i}(\omega_{qg} - \omega_P)\sigma_P)]\mathrm{e}^{-\mathrm{i}\omega_{qg}\bar{t}}. \tag{3.34d}$$

where we have defined $\bar{t} = t - t_{P'}$.[6] Also, in taking the Hermitian conjugate of $|\Psi_{+P-P}(t)\rangle$ in equation (3.30*b*) to substitute in equation (3.34), we have used the fact that $\mathrm{erf}(-\mathrm{i}x) = -\mathrm{erf}(\mathrm{i}x)$ for a real valued x. Terms such as $\langle\Psi_{+P+P'}(t)|\boldsymbol{\mu}|\Psi_{+P+P'}(t)\rangle$ do not contribute because they are fourth order in η and vanish anyway due to the form of the dipole operator (equation (2.9)) and the RWA. Hence, the next nonnegligible order of nonlinear polarization (not shown) would be fifth order in η.

The structure of equations (3.34*a*)–(3.34*d*) follows relatively simple patterns and can almost be trivially read out without much calculation by introducing the mnemonics of Double-Sided-Feynman-Diagrams (DS-FDs) in Liouville space[7] (see figure 3.2(*b*)). In fact, these equations can be obtained by taking the wavefunction diagrams from figure 3.2(*a*) and putting them next to their complex conjugates in a way that they satisfy the phase-matching condition $\mathbf{k}_{P'}$. In this way, the first DS-FD in figure 3.2(*b*) corresponds to the linear polarization due to P' (equation (3.2)). The nonlinear polarization diagrams are associated with processes of excited-state absorption (ESA) (equation (3.34*a*)), stimulated emission (SE) (equation (3.34*b*), and ground-state bleaching (equations (3.34*c*) for left-bottom and (3.34*d*) for right-bottom diagram)). The physical meaning of these terms will be clarified in chapter 4.

For a more comprehensive exposition of diagrammatics and spectroscopy, we refer the reader to [2, 6]. Here are the basic rules to construct DS-FDs:

1. Time in the DS-FD flows from bottom to top.
2. The ket and bra spaces are on the left and right, respectively.

[6] In the photon-echo literature, $\bar{t}$ is often called the *echo time*, and its symbol is t instead of $\bar{t}$. We use $\bar{t}$ in order to avoid confusion with t, which is the generic time variable in this book.

[7] Wavefunctions live in Hilbert space, whereas density matrices are in Liouville space.

3. The blue arrows indicate pulse perturbations (excitations denoted by arrows into the diagram, de-excitations denoted by the opposite). Each perturbation on the ket/bra contributes a factor of (+i)/(−i).
4. The red arrow indicates measurement of the dipole moment; do not confuse it with a perturbation. No factor of (+i)/(−i) is involved here.

As a simple illustration, consider the ESA pathway corresponding to equation (3.34*a*). Let us first do a simple check. The ket $|\Psi_{+P+P'}(t)\rangle$, associated with the phase $\mathrm{e}^{\mathrm{i}(\mathbf{k}_P+\mathbf{k}_{P'})\cdot\mathbf{r}+\mathrm{i}(\phi_P+\phi_{P'})}$, is described by equation (3.29*a*), whereas the bra $\langle\Psi_{+P}(t)|$, linked with $\left(\mathrm{e}^{\mathrm{i}\mathbf{k}_P\cdot\mathbf{r}+\mathrm{i}\phi_P}\right)^* = \mathrm{e}^{-\mathrm{i}\mathbf{k}_P\cdot\mathbf{r}-\mathrm{i}\phi_P}$, is given by the complex conjugate of equation (3.19*a*) for $n = P$. Both wavefunctions are shown in figure 3.2(*a*). Hence, the phase associated with the dipole matrix element of this particular DS-FD, $\langle\Psi_{+P}(t)|\boldsymbol{\mu}|\Psi_{+P+P'}(t)\rangle$, is $\mathrm{e}^{\mathrm{i}(\mathbf{k}_P+\mathbf{k}_{P'}-\mathbf{k}_P)\cdot\mathbf{r}+\mathrm{i}(\phi_P+\phi_{P'}-\phi_P)} = \mathrm{e}^{\mathrm{i}\mathbf{k}_{P'}\cdot\mathbf{r}+\mathrm{i}\phi_P}$, which is the correct phase for it to contribute to $\mathbf{P}_{\mathbf{k}_{P'}}(t)$.

We now describe the anatomy of $\langle\Psi_{+P}(t)|\boldsymbol{\mu}|\Psi_{+P+P'}(t)\rangle$ in more detail,

$$\langle\Psi_{+P}(t)|\boldsymbol{\mu}|\Psi_{+P+P'}(t)\rangle = \mathrm{i}\sum_{p,q=\alpha,\beta} \underbrace{\boldsymbol{\mu}_{pf}}_{\text{final dipole element}} \left(\underbrace{\Omega_{fq}^{P'}\Omega_{qg}^{P}}_{\text{perturbations on the ket}}\right)\left(\underbrace{\Omega_{gp}^{\overline{P}}}_{\text{perturbations on the bra}}\right)$$
$$\times\left(\underbrace{\mathrm{e}^{-\mathrm{i}\omega_{fp}\bar{t}}}_{\text{evolution between } t_{P'} \text{ and final } t}\right)\left(\underbrace{\mathrm{e}^{-\mathrm{i}\omega_{qp}T}}_{\text{evolution between } t_P \text{ and } t_{P'}}\right). \tag{3.35}$$

The corresponding transition amplitudes due to the perturbations in the ket and in the bra, can be read off from the corresponding DS-FD, starting with the initial state $|g\rangle\langle g|$ in Liouville space,

$$\overbrace{\underbrace{|f\rangle \xleftarrow{\Omega_{fq}^{P'}} |q\rangle \xleftarrow{\Omega_{qg}^{P}} |g\rangle}_{\text{ket}}\underbrace{\langle g| \xrightarrow{\Omega_{gp}^{\overline{P}}} \langle p|}_{\text{bra}}}^{\text{Final dipole element } \langle p|\boldsymbol{\mu}|f\rangle=\boldsymbol{\mu}_{pf} \text{ connecting the ket and bra sides at the end time } t} \tag{3.36}$$

The excitations on the ket proceed via ε_n terms, whereas those of the bra happen through ε_n^*, consistent with the RWA. De-excitations in each space follow the complementary rule. The final state in the ket is $|f\rangle$ and the one in the bra is $\langle p|$ for $p = \alpha$, β. Upon final measurement of the dipole $\boldsymbol{\mu}$, we obtain the dipole element $\langle p|\boldsymbol{\mu}|f\rangle = \boldsymbol{\mu}_{pf}$ multiplying the perturbations. We have placed it on the ket side, pointing out of the diagram, with the idea that a 'de-excitation' of the ket $|f\rangle$ via $\boldsymbol{\mu}$ will couple to the bra $\langle p|$ via $\boldsymbol{\mu}_{pf}$, but we could have placed it on the bra side pointing into the diagram conveying the same idea, that an 'excitation' of $\langle p|$ via $\boldsymbol{\mu}$ couples to $|f\rangle$.

By identifying the populations or coherences at each time interval in the DS-FD, we can easily write their time evolution: at the beginning, $|g\rangle\langle g|$ evolves trivially from 0 to t_P as $\mathrm{e}^{-\mathrm{i}\omega_{gg}(t_P-0)} = 1$. Upon perturbations in the bra and in the ket, both due to P, the state $|q\rangle\langle p|$ is generated, which evolves as $\mathrm{e}^{-\mathrm{i}\omega_{qp}(t_{P'}-t_P)}$. Finally, a perturbation on the ket via P' creates $|f\rangle\langle p|$, which evolves from $t_{P'}$ to t as $\mathrm{e}^{-\mathrm{i}\omega_{fp}(t-t_{P'})}$.

Recall that, since transitions between the GSM, SEM and DEM are assumed to be slow in the absence of light, we are interested in exploring the evolution of the SEM coherence or population term $|q\rangle\langle p|$ during the waiting time. In this simple model, that evolution is simply $e^{-i\omega_{qp}T}$.

3.2.3 Time and energy scales in the model

We finish our discussion of this idealized model by analyzing the approximations from the last section in light of typical values for the carrier frequencies and pulse widths. Let $\omega_P = \omega_{P'} = 12\,800\ \text{cm}^{-1}$ and $\sigma_P = \sigma_{P'} = 20$ fs, which are typical energy and timescales in experiments [5].

In our example, the bandwidth of the pulses is given by[8],

$$\sigma_P^{-1} = \left(\frac{1}{2\pi}\right)\left[\left(\frac{1}{20\ \text{fs}}\right)\left(\frac{10^{15}\ \text{fs}}{1\ \text{s}}\right)\left(\frac{1\ \text{s}}{3\times 10^{10}\ \text{cm}}\right)\right] = 265\ \text{cm}^{-1}.$$

We chose the center pulse frequency to be $\omega_n = \omega_P = \omega_{P'} = 12\,800\ \text{cm}^{-1}$, so $\omega_{\beta g} - \omega_n = 145\ \text{cm}^{-1}$ and $\omega_{\alpha g} - \omega_n = -145\ \text{cm}^{-1}$. Both of the transition energies are within one unit of σ_P^{-1}, so the envelope of $\tilde{\varepsilon}_n(\omega)$ is still considerable at those transition frequencies, and the pulses are effectively broadband. In fact, we have $\tilde{\varepsilon}_n(\omega_{\alpha g}) = \tilde{\varepsilon}_n(\omega_{f\beta}) = \tilde{\varepsilon}_n(\omega_{\beta g}) = \tilde{\varepsilon}_n(\omega_{f\alpha}) = 0.86$. On the other hand, $\tilde{\varepsilon}_n(-\omega_{\alpha g}),\ \tilde{\varepsilon}_n(-\omega_{f\beta}),\ \tilde{\varepsilon}_n(-\omega_{\beta g}),\ \tilde{\varepsilon}_n(-\omega_{f\alpha}) < e^{-1000}$, which validates the RWA. That is, the pulses are broadband compared to the bandwidth of each excitation manifold, $|\omega_\alpha - \omega_\beta| < \sigma$, but not broadband when compared with the bandwidth of the two optical gaps, $|\omega_{\alpha g} + \omega_n| > \sigma$.

Another way to frame these statements is to think about what $\gg$ means in assumptions 1 to 3 (equations (3.9)–(3.11)), which validate our expressions from parts 2 and 3. We note that the pulses peak at $t = t_n$ and are effectively <1% of their maximum amplitude for $t \notin [t_n - 3\sigma_n,\ t_n + 3\sigma_n]$, where $3\sigma_n = 60$ fs. On the other hand, there are two timescales for dynamics. One is given by the timescale of the evolution of the coherent superpositions between states in different excitation manifolds (sometimes known as optical coherences, which oscillate at the frequencies $\pm\omega_{\alpha g} = \pm\omega_{f\beta}$ and $\pm\omega_{\beta g} = \pm\omega_{f\alpha}$), and is on the order of the optical period $\frac{2\pi}{\omega_n} = 2.60$ fs. The other timescale is the one given by coherent superpositions

[8] Beware of the factor of $(\frac{1}{2\pi})$ in the calculation, which comes from our convention in this book to report energies in cm^{-1} and carrier frequencies of the pulses in terms of ordinary (instead of angular) frequencies. If confused, work with units of time and angular frequency, fs and fs^{-1}. As an example, the period T it takes for the phase of light at frequency $12\,800\ \text{cm}^{-1}$ to go through 2π radians is,

$$T = \underbrace{\left(\frac{1}{12\,800\ \text{cm}^{-1}}\right)}_{\text{Wavelength}}\underbrace{\left(\frac{10^5\ \text{fs}}{3\ \text{cm}}\right)}_{c^{-1}} = 2.60\ \text{fs},$$

which in turn yields a time of 2.60 fs/2π = 0.41 fs for the phase to evolve one radian.

in the SEM, $\frac{2\pi}{\omega_{\beta\alpha}} = 115$ fs. Hence, the pulses are short only compared to the dynamics within an excitation manifold, in our case, the SEM. This is just a time-domain perspective of the frequency-domain conclusions from the previous paragraph.

These numerical values validate $\mathbf{P}_{\mathbf{k}_{p'}}(t)$ as consistent with the picture of ultrafast spectroscopy that we presented in chapter 2: short pulses induce transitions and allow us to watch the dynamics in the SEM in real time between the pulses. Finally, we note that the time dependence of $\mathbf{P}_{\mathbf{k}_{p'}}(t)$ is almost trivial owing to the simplicity of the toy model we are exploring. This will change when we consider decoherence of the reduced electronic dynamics caused by the nuclei.

Equation (3.5) for the polarization grating $\mathbf{P}(\mathbf{r}, t)$ is expressed in terms of the reduced electronic state $\rho(\mathbf{r}, t)$ of the molecule located at $\mathbf{r}$. Throughout this section we have assumed that the calculation of this reduced state can be achieved by focusing on each molecule independently while ignoring the rest of the chromophores in the ensemble. Appendix C examines this assumption in more detail and shows that a more formal calculation considering the 'many-molecule' quantum state of the ensemble yields the same results as the 'independent-molecule' one.

3.3 Measuring the signal: connecting induced polarization to experimental results

In the previous section, we computed the polarization induced in a sample by a series of short laser pulses. In general, this polarization is indirectly measured by monitoring the energy lost by one of the pulses upon transmission through the sample. Let us develop this idea more carefully by considering the calculation of the rate of energy absorbed from light by a molecule located at position $\mathbf{r}$. Using equation (3.1) [6, 8],

$$\begin{aligned}
\frac{\mathrm{d}E(\mathbf{r},t)}{\mathrm{d}t} &\equiv \frac{\mathrm{d}}{\mathrm{d}t}\,\mathrm{Tr}\left[H(\mathbf{r},t)\rho(\mathbf{r},t)\right] \\
&= \underbrace{\mathrm{Tr}\left[\frac{\mathrm{d}H(\mathbf{r},t)}{\mathrm{d}t}\rho(\mathbf{r},t)\right]}_{=-\mathrm{Tr}\left[\boldsymbol{\mu}\cdot\dot{\boldsymbol{\varepsilon}}(\mathbf{r},t)\rho(\mathbf{r},t)\right]} + \underbrace{\mathrm{Tr}\left[H(\mathbf{r},t)\frac{\mathrm{d}}{\mathrm{d}t}\rho(\mathbf{r},t)\right]}_{=-\mathrm{i}\,\mathrm{Tr}\left[H[H,\rho]\right]=0} \\
&= \sum_{n=1}^{N_{\mathrm{pulses}}}\frac{\mathrm{d}E_n}{\mathrm{d}t},
\end{aligned} \tag{3.37}$$

where in the second line we have used the fact that the molecular Hamiltonian H_0 is not time-dependent, introduced $\dot{\boldsymbol{\varepsilon}}(\mathbf{r},t) = \frac{\mathrm{d}}{\mathrm{d}t}\boldsymbol{\varepsilon}(\mathbf{r},t)$, and exploited Liouville's equation for the density matrix $\rho(\mathbf{r}, t)$ of the molecule at $\mathbf{r}$. The third line establishes a partition of the rate in terms of the various incoming pulses. In particular, let us single out the last N_{pulses} field and name it the *local oscillator* (LO),

$\varepsilon_{LO}(t - t_p)e^{i\mathbf{k}_{LO}\cdot\mathbf{r}+i\phi_{LO}}\mathbf{e}_{LO}$ + c.c., which we shall regard as a reference pulse. Following equation (3.37),

$$\begin{aligned}\frac{dE_{LO}(\mathbf{r},t)}{dt} &= \text{Tr}\{-\boldsymbol{\mu}\cdot[\dot{\varepsilon}_{LO}(t-t_{LO})e^{i\mathbf{k}_{LO}\cdot\mathbf{r}+i\phi_{LO}}\mathbf{e}_{LO}+\text{c.c.}]\rho(\mathbf{r},t)\}\\ &= -[\dot{\varepsilon}_{LO}(t-t_{LO})e^{i\mathbf{k}_{LO}\cdot\mathbf{r}+i\phi_{LO}}\mathbf{e}_{LO}+\text{c.c.}]\cdot\mathbf{P}(\mathbf{r},t)\\ &= -2\Re\sum_s e^{i(\mathbf{k}_s-\mathbf{k}_{LO})\cdot\mathbf{r}+i(\phi_s-\phi_{LO})}\dot{\varepsilon}^*_{LO}(t-t_{LO})\mathbf{e}^*_{LO}\cdot\mathbf{P}_{\mathbf{k}_s}(t), \end{aligned} \tag{3.38}$$

where we have used the definition and decomposition of the polarization $\mathbf{P}(\mathbf{r}, t)$ from equation (3.5). We can also calculate the *total* energy of the LO absorbed by that particular molecule by time integrating equation (3.38),

$$\Delta E_{LO}(\mathbf{r}) = -2\Re\sum_s\int_{-\infty}^{\infty}dt' e^{i(\mathbf{k}_s-\mathbf{k}_{LO})\cdot\mathbf{r}+i(\phi_s-\phi_{LO})}\dot{\varepsilon}^*_{LO}(t'-t_{LO})\mathbf{e}^*_{LO}\cdot\mathbf{P}_{\mathbf{k}_s}(t'). \tag{3.39}$$

Equations (3.38) and (3.39) highlight the fact that molecules located at different positions $\mathbf{r}$ will in general absorb different amounts of energy from the LO depending on the values of the $e^{i(\mathbf{k}_s-\mathbf{k}_{LO})\cdot\mathbf{r}}$ phase factors.

Example 3. Energy vs number of photons absorbed by the sample

The energy described in equation (3.39) can also be partitioned in the frequency domain,

$$\Delta E_{LO}(\mathbf{r}) = \int_{-\infty}^{\infty}\Delta E_{LO}(\mathbf{r},\omega)d\omega, \tag{3.40}$$

$$\Delta E_{LO}(\mathbf{r},\omega) = \omega\Delta N_{LO}(\mathbf{r},\omega), \tag{3.41}$$

where $\Delta E_{LO}(\mathbf{r}, \omega)$ is the total energy lost by the LO due to absorption of $N_{LO}(\mathbf{r}, \omega)$ photons of frequency ω by the molecule at position $\mathbf{r}$.

1. Derive an expression for $N_{LO}(\mathbf{r}, \omega)$ in terms of the appropriate Fourier transforms of the pulse and the polarization. Utilize the convention for Fourier transforms as in equation (3.21), namely,

$$\tilde{f}(\omega) = \int_{-\infty}^{\infty}dt f(t)e^{i\omega t}\leftrightarrow f(t) = \frac{1}{2\pi}\int_{-\infty}^{\infty}d\omega\tilde{f}(\omega)e^{-i\omega t}. \tag{3.42}$$

2. Let $N_{LO}(\mathbf{r}) = \int_{-\infty}^{\infty}N_{LO}(\mathbf{r},\omega)d\omega$ be a frequency integrated count of the number of photons. Express $N_{LO}(\mathbf{r})$ as an integral in the time domain.

Solution

1. For this example, we closely follow the formalism developed in [8], chapter 13. Equation (3.39) contains an inner product in the time domain, which can be translated to another inner product in the frequency domain. From equation (3.42), it can be easily checked that an inner product in the time domain is equivalent to an inner product in the frequency domain,

$$\int_{-\infty}^{\infty} dt f^*(t)g(t) = \frac{1}{2\pi}\int_{-\infty}^{\infty} d\omega \tilde{f}^*(\omega)\tilde{g}(\omega). \tag{3.43}$$

The pulse and its time derivative transform as,

$$\int_{-\infty}^{\infty} dt\, \varepsilon_{\mathrm{LO}}(t - t_{\mathrm{LO}})e^{i\omega t} = e^{i\omega t_{\mathrm{LO}}}\tilde{\varepsilon}_{\mathrm{LO}}(\omega), \tag{3.44}$$

$$\int_{-\infty}^{\infty} dt\, \dot{\varepsilon}_{\mathrm{LO}}(t - t_{\mathrm{LO}})e^{i\omega t} = -i\omega e^{i\omega t_{\mathrm{LO}}}\tilde{\varepsilon}_{\mathrm{LO}}(\omega), \tag{3.45}$$

whereby equation (3.39) can be rewritten as,

$$\Delta E_{\mathrm{LO}}(\mathbf{r}) = \frac{1}{\pi}\Im\int_{-\infty}^{\infty} d\omega\, \omega\sum_s e^{i(\mathbf{k}_s-\mathbf{k}_{\mathrm{LO}})\cdot\mathbf{r}+i(\phi_s-\phi_{\mathrm{LO}})}e^{-i\omega t_{\mathrm{LO}}}\tilde{\varepsilon}^*_{\mathrm{LO}}(\omega)\mathbf{e}^*_{\mathrm{LO}}\cdot\tilde{\mathbf{P}}_{\mathbf{k}_s}(\omega). \tag{3.46}$$

Comparing with equation (3.41),

$$N_{\mathrm{LO}}(\mathbf{r}, \omega) = \frac{1}{\pi}\Im\sum_s e^{i(\mathbf{k}_s-\mathbf{k}_{\mathrm{LO}})\cdot\mathbf{r}+i(\phi_s-\phi_{\mathrm{LO}})}e^{-i\omega t_{\mathrm{LO}}}\tilde{\varepsilon}^*_{\mathrm{LO}}(\omega)\mathbf{e}^*_{\mathrm{LO}}\cdot\tilde{\mathbf{P}}_{\mathbf{k}_s}(\omega). \tag{3.47}$$

2. By using the inner product property of equation (3.43) again,

$$N_{\mathrm{LO}}(\mathbf{r}) = 2\Im\sum_s\int_{-\infty}^{\infty} dt e^{i(\mathbf{k}_s-\mathbf{k}_{\mathrm{LO}})\cdot\mathbf{r}+i(\phi_s-\phi_{\mathrm{LO}})}\varepsilon^*_{\mathrm{LO}}(t - t_{\mathrm{LO}})\mathbf{e}^*_{\mathrm{LO}}\cdot\mathbf{P}_{\mathbf{k}_s}(t), \tag{3.48}$$

so that the total number of lost photons by a particular molecule can be thought of as the integral of the time overlap between the LO and the polarization.

Finally, even though the LO is typically broadband, it is still centered about a particular optical frequency (see discussion about the RWA in section 3.2). From equation (3.4), this means that the main contribution to the time derivative of the pulse is the one associated with its carrier frequency and not its bandwidth, $\dot{\varepsilon}_{\mathrm{LO}}(t - t_{\mathrm{LO}}) \approx -i\omega_{\mathrm{LO}}\varepsilon_{\mathrm{LO}}(t - t_{\mathrm{LO}})$. Inserting this approximation in equation (3.39), we get,

$$\Delta E_{\mathrm{LO}}(\mathbf{r}) \approx \omega_{\mathrm{LO}}N_{\mathrm{LO}}(\mathbf{r}). \tag{3.49}$$

This is a very intuitive result, and it can also be thought of as assuming $\omega \approx \omega_{\mathrm{LO}}$ in equation (3.40) and integrating it as in equation (3.39). It simply indicates the approximate statement that the absorbed energy is proportional to the number of absorbed photons, which all have roughly the same energy. Importantly, when the bandwidth of the pulse is very large, this approximation is expected to fail; in such case, equation (3.46) is preferred to equation (3.49).

Several experimental measurements in ensembles of many molecules are possible; for instance, a total count of frequency-resolved or frequency-integrated photons absorbed by the material. This amounts to adding up the signals corresponding to equations (3.39), (3.41), (3.47), or (3.48) across all the **r** values in the ensemble.

Therefore, we need to obtain the total photon loss as an integral over the photon loss per molecule in the sample. Assuming N molecules are randomly distributed throughout a volume V, we can convert the discrete sum into an integral $\sum_{\mathbf{r}} \rightarrow \frac{N}{V}\int d^3r$, giving,

$$\begin{aligned}\int d^3r e^{i(\mathbf{k}_s-\mathbf{k}_{LO})\cdot\mathbf{r}} &= \lim_{L_i\to\infty}\int_0^{L_x} dx \int_0^{L_y} dy \int_0^{L_z} dz\, e^{i(\mathbf{k}_s-\mathbf{k}_{LO})\cdot\mathbf{r}} \\ &= (2\pi)^3\delta^3(\mathbf{k}_{LO}-\mathbf{k}_s), \end{aligned} \tag{3.50}$$

where we have denoted $\mathbf{L} = (L_x, L_y, L_z)$ and $\delta^3(\mathbf{k})$ is the three-dimensional Dirac delta function. The integral has been carried out in a large box compared to the characteristic wavelength of the pulses (a typical laser spot size is about 10 μm and L_i should be less than that, see appendix A). Of particular interest in this book will be the *frequency-integrated photon loss* of the LO, which by virtue of equations (3.48) and (3.50), is given by,

$$\begin{aligned}\Delta N_{LO} &= \sum_{\mathbf{r}} N_{LO}(\mathbf{r}) \\ &\approx \frac{N}{V}(2\pi)^3\delta^3(\mathbf{k}_{LO}-\mathbf{k}_s)S_s, \end{aligned} \tag{3.51}$$

where N is the number of molecules in the array and V is the volume, and we have defined the *signal* corresponding to the average number of photons lost by the LO, per molecule, and in the $\mathbf{k}_s$ direction, as

$$S_s \equiv 2\Im\int_{-\infty}^{\infty} dt' \varepsilon^*_{LO}(t'-t_{LO})e^{-i\phi_{LO}}\mathbf{e}^*_{LO}\cdot\mathbf{P}_{\mathbf{k}_s}(t')e^{i\phi_s}. \tag{3.52}$$

Equations (3.51)–(3.52) are the main results of this chapter. The delta-function $\delta^3(\mathbf{k}_{LO}-\mathbf{k}_s)$ shows that energy can only be transmitted or absorbed by the LO from the Fourier mode of the polarization grating associated with the same wavevector as the pulse. This condition is often called *phase matching*, as it relies on the condition $e^{i\mathbf{k}_{LO}\cdot\mathbf{r}} \approx e^{i\mathbf{k}_s\cdot\mathbf{r}}$. The other Fourier components interfere destructively with the LO, and no energy is retrieved from them. This is the same as to say that for polarization components that do not match the LO wavevector, the molecules will absorb (emit) different amounts of energy from (to) the LO as a function of spatial location. After averaging over the whole ensemble, the total energy gained (lost) for these components will be zero. Therefore, only finite exchange of energy will occur for components of the polarization grating that match the LO wavevector. Notice that this result holds regardless of whether the LO participates in creating the polarization component $\mathbf{P}_{\mathbf{k}_s}(t)$. This observation will be made clear in the next chapters[9].

[9] The transmission of the LO *through* the sample is not necessary for the phase matching result to hold. It suffices that it mixes with the weak radiation emitted by the polarization grating outside the sample. Interested readers can consult [6, 7].

Throughout this book, we will focus on various *frequency-integrated* spectroscopic signals S_s based on total count of photons, irrespective of color. Appendix D describes additional spectroscopic possibilities obtained from dissecting the photon-counts by frequency using a spectrophotometer. This enterprise is known as *frequency-resolved spectroscopy*, the most common examples of it being the standard linear absorption spectrum (example 12) and the frequency-resolved pump–probe spectroscopy (example 14). These two types of spectroscopies are obviously related, and the reader should be aware of the possibility that a particular instance of frequency-integrated nonlinear spectrum may yield the same information as some other frequency-resolved nonlinear spectrum. For clarity and brevity, we will limit our discussion to the former. However, the reader is invited to appendix D to explore those connections, which continue to be the subject of active research.

Bibliography

[1] Biggs J D and Cina J A 2009 Using wave-packet interferometry to monitor the external vibrational control of electronic excitation transfer *J. Chem. Phys.* **131** 224101

[2] Cho M 2009 *Two Dimensional Optical Spectroscopy* (Boca Raton, FL: CRC Press)

[3] Cohen-Tannoudji C, Diu B and Laloe F 1977 *Quantum Mechanics* vol 1 (New York: Wiley-Interscience)

[4] Kjellberg P and Pullerits T 2006 Three-pulse photon echo of an excitonic dimer modeled via Redfield theory *J. Chem. Phys.* **124** 024106

[5] Lee H, Cheng Y C and Fleming G R 2007 Coherence dynamics in photosynthesis: protein protection of excitonic coherence *Science* **316** 1462–5

[6] Mukamel S 1995 *Principles of Nonlinear Optical Spectroscopy* (Oxford: Oxford University Press)

[7] Rohrdanz M A and Cina J A 2006 Probing intermolecular communication via lattice phonons with time-resolved coherent anti-Stokes Raman scattering *Mol. Phys.* **104** 1161–78

[8] Tannor D J 2007 *Introduction to Quantum Mechanics: A Time Dependent Approach* (Mill Valley, CA: University Science Books)

[9] Yuen-Zhou J, Krich J J, Mohseni M and Aspuru-Guzik A 2011 Quantum state and process tomography of energy transfer systems via ultrafast spectroscopy *Proc. Natl Acad. Sci. USA* **108** 17615–20

Chapter 4

Interaction of light pulses with ensembles of chromophores: wavepackets

The spectroscopic observable given by equation (3.52) can be readily calculated once the components $\mathbf{P}_{\mathbf{k}_s}(t)$ of the polarization in the expansion of equation (3.5) are known. In order to develop our intuition further, we will reexpress these observables in terms of overlaps between wavepackets created by the different pulses at different points in time.

4.1 Linear absorption spectroscopy

The simplest example of the general theory presented above consists of linear absorption spectroscopy, as depicted in figure 4.1. A pump pulse P travels with wavevector $\mathbf{k}_p$ and transfers amplitude from the ground electronic state to the dipole-allowed excited states. At $t = 0$, we consider that the chromophore at each position $\mathbf{r}$ is in a particular vibronic eigenstate $|\Psi_0\rangle$ of H_{GSM}, so that it undergoes trivial time evolution in the absence of pulses, $|\Psi_0(t)\rangle = \mathrm{e}^{-\mathrm{i}E_0 t}|\Psi_0\rangle$ (equation (3.18b)). The result for a thermal mixture of initial states in the GSM is readily obtained from the results below by incoherently (classically) averaging over the possible initial states.

The total wavefunction $|\Psi(\mathbf{r}, t)\rangle$ for the molecule at location $\mathbf{r}$ after the pulse acts is, to first order in η, given by,

$$\begin{aligned} |\Psi(\mathbf{r}, t)\rangle &= |\Psi_0(t)\rangle + \mathrm{e}^{\mathrm{i}\mathbf{k}_P\cdot\mathbf{r}+\mathrm{i}\phi_P}|\Psi_{+P}(t)\rangle + \mathrm{e}^{-\mathrm{i}\mathbf{k}_P\cdot\mathbf{r}-\mathrm{i}\phi_P}|\Psi_{-P}(t)\rangle \\ &\approx |\Psi_0(t)\rangle + \mathrm{e}^{\mathrm{i}\mathbf{k}_P\cdot\mathbf{r}+\mathrm{i}\phi_P}|\Psi_{+P}(t)\rangle, \end{aligned} \tag{4.1}$$

where $|\Psi_{+P}(t)\rangle$ and $|\Psi_{-P}(t)\rangle$ are given by equations (3.19a)[1] and (3.24), where $|\Psi_{-P}(t)\rangle$ vanishes in the RWA. We note that even though in section 3.2 we treated an isolated electronic system $\mathscr{S}$, the wavefunction $|\Psi(\mathbf{r}, t)\rangle$ here is, in general, a wavepacket in the entire $\mathscr{S}$ (electronic) and $\mathscr{B}$ (vibrational) space.

[1] Note that equation (3.19a) is general, but equations (3.19b) and (3.19c) do not apply here as they are specific to the vibrationless model of section 3.2.

doi:10.1088/978-0-750-31062-8ch4

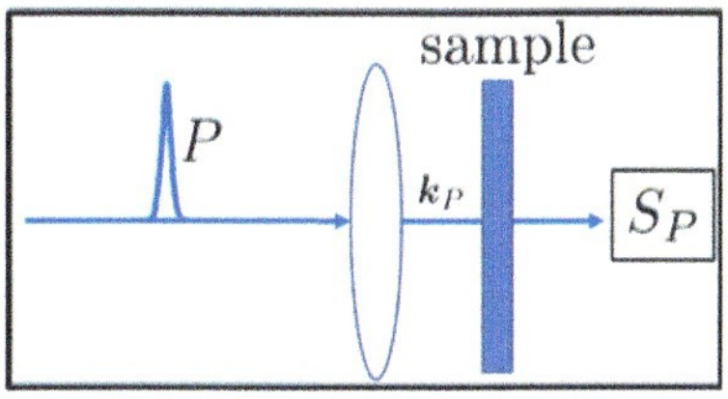

Figure 4.1. Linear absorption spectroscopy. A pulse P with wavevector $\mathbf{k}_p$ passes through a lens and then through a sample with chromophores. The sample absorbs an average number of S_P photons per chromophore.

It will be convenient to define the first-order wavepacket created by P on the molecule at $\mathbf{r} = 0$,

$$\begin{aligned} |\Psi_P(t)\rangle &\equiv -\mathrm{i}\int_{-\infty}^{t} \mathrm{d}t' \mathrm{e}^{-\mathrm{i}H_0(t-t')}(-\boldsymbol{\mu}\cdot\mathbf{e}_P)[\varepsilon_P(t'-t_P)+\text{c.c.}]|\Psi_0(t')\rangle \\ &= |\Psi_{+P}(t)\rangle + |\Psi_{-P}(t)\rangle, \\ &\approx |\Psi_{+P}(t)\rangle, \end{aligned} \tag{4.2}$$

where the RWA (equation (3.25)) allows us to drop the contribution of $|\Psi_{-P}(t)\rangle$. Although seemingly redundant, this definition will help us organize the various terms that appear in the perturbation expansions corresponding to more than one pulse (see section 4.2).

The physical interpretation of these equations is the following: the stationary wavepacket evolves (trivially) in the ground electronic surface up to time t', it gets promoted to the dipole-allowed excited state(s) with a probability amplitude $-\boldsymbol{\mu}\cdot\mathbf{e}_P\varepsilon_P(t'-t_P)$, and then continues evolving via the free Hamiltonian H_0 in the excited state(s) from t' to time t. The resulting wavepacket is the superposition of the different wavepackets promoted to the excited state(s) at all the different times t' allowed by the Gaussian envelope $\varepsilon_P(t'-t_P)$ and the upper limit of the integral at $t' = t$.

We shall also define the asymptotic wavepacket,

$$|\Psi_P\rangle \equiv \lim_{t\to\infty} \mathrm{e}^{\mathrm{i}H_0 t}|\Psi_P(t)\rangle \tag{4.3}$$

as the wavepacket created by P after the *full* pulse P has acted, where the operator $\mathrm{e}^{\mathrm{i}H_0 t}$ guarantees a well-defined phase. The induced polarization grating, to first order in η, has the form $\mathbf{P}(\mathbf{r}, t) = \mathbf{P}_{\mathbf{k}_P}(t)\mathrm{e}^{\mathrm{i}\mathbf{k}_P\cdot\mathbf{r}+\mathrm{i}\phi_P} + \text{c.c.}$, where

$$\mathbf{P}_{\mathbf{k}_P}(t) = \langle\Psi_0(t)|\boldsymbol{\mu}|\Psi_P(t)\rangle. \tag{4.4}$$

If we are interested in the energy of P absorbed by the sample, we can regard P as both the trigger of the polarization and as the LO, and using equation (3.52), write,

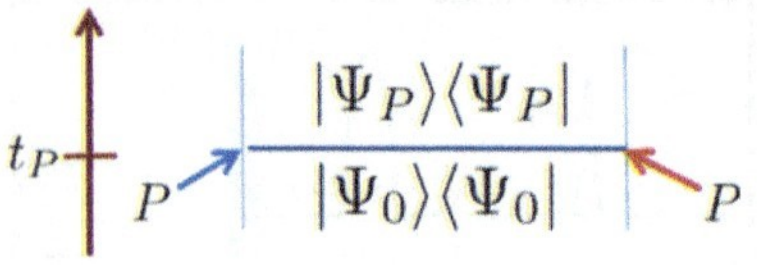

Figure 4.2. DS-FD for linear absorption.

$$
\begin{aligned}
S_P &= 2\Im\int_{-\infty}^{\infty} dt'\varepsilon_P^*(t'-t_P)e^{-i\phi_P}\mathbf{e}_P^*\cdot\mathbf{P}_{\mathbf{k}_P}(t')e^{i\phi_P}\\
&= 2\Im i\int_{-\infty}^{\infty} dt'\underbrace{\langle\Psi_0(t')|(-i)\varepsilon_P^*(t'-t_p)\boldsymbol{\mu}\cdot\mathbf{e}_P^*}_{=\partial_{t'}\langle\Psi_P(t')|}|\Psi_P(t')\rangle\\
&= 2\Im i\int_{-\infty}^{\infty} dt'[\partial_{t'}\langle\Psi_P(t')|]|\Psi_P(t')\rangle\\
&= \int_{-\infty}^{\infty} dt'\partial_{t'}[\langle\Psi_P(t')|\Psi_P(t')\rangle]\\
&= \langle\Psi_P|\Psi_P\rangle.
\end{aligned}
\tag{4.5}
$$

In the second line we have identified the bra piece as the time derivative of equation (4.2). From the third to the fourth line we notice that the integrand can be written as the total derivative of the overlap $\langle\Psi_P(t')|\Psi_P(t')\rangle$. Finally, from the fourth to the fifth line, we have evaluated the result of the integral at the upper and lower limits, $\lim_{t'\to\infty}\langle\Psi_P(t')|\Psi_P(t')\rangle = \langle\Psi_P|\Psi_P\rangle$ and $\lim_{t'\to-\infty}\langle\Psi_P(t')|\Psi_P(t')\rangle = 0$. Note that the phases ϕ_P do not play a role in the observable of interest. Figure 4.2 shows the corresponding DS-FD for this experiment.

The meaning of equation (4.5) is quite simple: the number of photons lost by P at the end of the absorption experiment is proportional to the sum of all the populations created on different excited states $\langle\Psi_P|\Psi_P\rangle = \sum_n\langle\Psi_P|n\rangle\,\langle n|\Psi_P\rangle$ (for some basis $\{|n\rangle\}$). Wavepackets were introduced into spectroscopy by Heller, in a manifestly time-dependent approach to the subject [2, 7, 8]. We emphasize that this is a frequency-integrated result, where we are uniformly counting all the photons lost due to material absorption, regardless of their color. As mentioned before, we refer the reader to appendix D and specifically to example 12 to learn about the *frequency-resolved* version of linear spectroscopy.

Example 4. Phase matching and absorption from a sample

In this example, we illustrate how phase-matching nontrivially affects the average absorption of molecules in an ensemble. Additionally, we provide a physically intuitive picture of what phase-matching entails from the perspective of each chromophore in the ensemble. The reader interested in moving to pump–probe spectroscopy may skip to section 4.2.

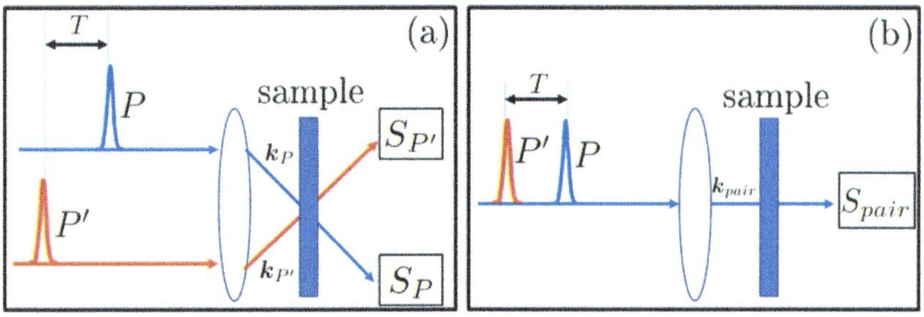

Figure 4.3. Linear absorption from a pulse pair in (a) noncollinear and (b) collinear setups. Phase-matching considerations are key to the distinction between these two scenarios.

Consider again an ensemble of vibrationless coupled dimers that interact with two pulses P and P'. Let the assumptions from equations (3.9)–(3.11) hold. In addition, let $\omega_P = \omega_{P'}$, $\sigma_P = \sigma_{P'}$, $\mathbf{e}_P = \mathbf{e}_{P'}$ and $\phi_P = \phi_{P'}$ so that the pulse ε_P is identical to $\varepsilon_{P'}$ except for the center times t_P and $t_{P'}$.

1. As usual, let $\mathbf{k}_P$ and $\mathbf{k}_{P'}$ be wavevectors pointing along different directions (figure 4.3(*a*)). Calculate S_P and $S_{P'}$, the total photon number loss for each direction, to first order in η.
2. Consider now a collinear setting. Define $\mathbf{k}_{\text{pair}} \equiv \mathbf{k}_P = \mathbf{k}_{P'}$ (figure 4.3(*b*)). What is the photon number loss S_{pair} from the two pulses *as a whole*? Repeat the analysis separately for each pulse.
3. Interpret the difference in the results from parts 1 and 2.

Solution

1. Using equations (3.19*a*), (4.2), (4.3), and (4.5),

$$
\begin{aligned}
|\Psi_P\rangle &= \lim_{t\to\infty} \mathrm{e}^{\mathrm{i}H_0 t} i\eta \sum_{q=\alpha,\beta} |q\rangle \Omega^P_{qg} \mathrm{e}^{-\mathrm{i}\omega_q(t-t_P)} \mathrm{e}^{-\mathrm{i}\omega_g t_P} \\
&= i\eta \sum_{q=\alpha,\beta} |q\rangle \Omega^P_{qg} \mathrm{e}^{\mathrm{i}\omega_q t_P} \mathrm{e}^{-\mathrm{i}\omega_g t_P},
\end{aligned}
\tag{4.6}
$$

$$
\begin{aligned}
S_P &= \langle \Psi_P | \Psi_P \rangle \\
&= \sum_{q=\alpha,\beta} |\Omega^P_{qg}|^2
\end{aligned}
\tag{4.7}
$$

Since $\tilde{\varepsilon}_{P'}(\omega) = \tilde{\varepsilon}_P(\omega)$ and $\mathbf{e}_P = \mathbf{e}_{P'}$, it follows that $\Omega^P_{qg} = \Omega^{P'}_{qg}$ (see equation (3.20)). Also, $|\Psi_{P'}\rangle$ is identical to $|\Psi_P\rangle$ except for the substitution $t_P \to t_{P'}$. Therefore, $S_{P'} = S_P$, i.e., the number of (asymptotically) absorbed photons is unaffected by the arrival time of the pulse, as expected.
2. For the collinear case, we rewrite equation (3.8) as,

$$
\boldsymbol{\varepsilon}(\mathbf{r}, t) = \underbrace{[\varepsilon_P(t - t_P) + \varepsilon_{P'}(t - t_{P'})\mathrm{e}^{\mathrm{i}(\phi_{P'} - \phi_P)}]}_{\equiv \varepsilon_{\text{pair}}(t - t_{\text{pair}})} \mathrm{e}^{\mathrm{i}\overbrace{\mathbf{k}_P}^{\equiv \mathbf{k}_{\text{pair}}} \cdot \mathbf{r} + \mathrm{i}\overbrace{\phi_P}^{\equiv \phi_{\text{pair}}}} \overbrace{\mathbf{e}_P}^{\equiv \mathbf{e}_{\text{pair}}} + \text{c.c.}, \tag{4.8}
$$

where we have defined $\varepsilon_{\text{pair}}$, ϕ_{pair}, $\mathbf{e}_{\text{pair}}$ and $t_{\text{pair}} = t_P$ to fit the mathematical form of a standard pulse ε_n (equation (3.3)). This allows us to directly apply equations (3.19*a*) and (4.3) to obtain the asymptotic wavepacket,

$$|\Psi_{\text{pair}}\rangle = \mathrm{i}\sum_{q=\alpha,\beta} |q\rangle\Omega_{qg}^{P}[\mathrm{e}^{\mathrm{i}\omega_q t_P}\mathrm{e}^{-\mathrm{i}\omega_g t_P} + \mathrm{e}^{\mathrm{i}\omega_q t_{P'}}\mathrm{e}^{-\mathrm{i}\omega_g t_{P'}}\mathrm{e}^{\mathrm{i}(\phi_{P'}-\phi_P)}] \tag{4.9}$$

The total photon loss from both pulses altogether is, by equation (4.5),

$$\begin{aligned} S_{\text{pair}} &= \langle\Psi_{\text{pair}}|\Psi_{\text{pair}}\rangle \\ &= \sum_{q=\alpha,\beta} |\Omega_{qg}^{P}|^2(2 + 2\cos(\omega_{qg}T + \phi_{P'} - \phi_P)), \end{aligned} \tag{4.10}$$

which clearly depends on the waiting time T. Comparing equation (4.10) with equations (4.6) and (4.7), we may rewrite it as,

$$S_{\text{pair}} = \langle\Psi_P|\Psi_P\rangle + \langle\Psi_{P'}|\Psi_{P'}\rangle + 2\Re\langle\Psi_{P'}|\Psi_P\rangle. \tag{4.11}$$

where the interference term is responsible for the T dependence.

Let us now consider the photon loss for each pulse $n = P, P'$. From equation (3.52),

$$\begin{aligned} S_n &= 2\Im\int_{-\infty}^{\infty} \mathrm{d}t'\varepsilon_n^*(t' - t_n)\mathrm{e}^{-\mathrm{i}\phi_n}\mathbf{e}_n^* \cdot [\mathbf{P}_{\mathbf{k}_P}(t')\mathrm{e}^{\mathrm{i}\phi_P} + \mathbf{P}_{\mathbf{k}_{P'}}(t')\mathrm{e}^{\mathrm{i}\phi_P}] \\ &= 2\Im\left\{\mathrm{i}\int_{-\infty}^{\infty} \mathrm{d}t'\mathrm{e}^{-\mathrm{i}\phi_n}\underbrace{\langle\Psi_0(t')|(-\mathrm{i})\varepsilon_n(t' - t_n)\boldsymbol{\mu}\cdot\mathbf{e}_n^*}_{=\partial_{t'}\langle\Psi_n(t')|}[|\Psi_P(t')\rangle\mathrm{e}^{\mathrm{i}\phi_P}\right. \\ &\qquad \left. + |\Psi_{P'}(t')\rangle\mathrm{e}^{\mathrm{i}\phi_{P'}}]\right\} \\ &= 2\Re\int_{-\infty}^{\infty} \mathrm{d}t'\{[\partial_{t'}\langle\Psi_n(t')|]|\Psi_P(t')\rangle\mathrm{e}^{\mathrm{i}(\phi_P-\phi_n)} \\ &\qquad + [\partial_{t'}\langle\Psi_n(t')|]|\Psi_{P'}(t')\rangle\mathrm{e}^{\mathrm{i}(\phi_{P'}-\phi_n)}\}, \end{aligned} \tag{4.12}$$

where we have proceeded as in equation (4.5).

For $n = P$, the second term in the last line of equation (4.12) is negligible because it is an interference term between ε_P and the polarization due to P', $\mathbf{P}_{\mathbf{k}_{P'}}$. However, P and P' are short pulses and there is a waiting time T between them which is much longer than their pulse widths. Hence, $\mathbf{P}_{\mathbf{k}_{P'}} \approx 0$ for the effective time window of ε_P, and the resulting integral vanishes, $\int_{-\infty}^{\infty}\mathrm{d}t'\varepsilon_P^*(t' - t_P)\mathbf{e}_P \cdot \mathbf{P}_{\mathbf{k}_{P'}}(t') \sim 0$.[2] Hence, only the first term survives as in equation (4.5), and

$$S_P = \langle\Psi_P|\Psi_P\rangle. \tag{4.13}$$

[2] It is important not to apply equation (3.19*a*) throughout for $\mathbf{P}_{\mathbf{k}_{P'}}(t')$ since it is only valid for $t \gg t_{P'}$.

For $n = P'$, the situation is different. Both terms in the last line of equation (4.12) are nonzero. The first term is an interference between $\varepsilon_{P'}$ and the polarization $\mathbf{P}_{\mathbf{k}_P}$, which is due to P. As opposed to the previous case, by the time P' arrives, there is already a polarization $\mathbf{P}_{\mathbf{k}_P}$ triggered beforehand by P, so the first term of equation (4.12) becomes,

$$2\Im \mathrm{i}\int_{-\infty}^{\infty} \mathrm{d}t' \mathrm{e}^{-\mathrm{i}\phi_{P'}} \langle \Psi_0(t')|(-\mathrm{i})\varepsilon_{P'}^*(t'-t_{P'})\boldsymbol{\mu}\cdot\mathbf{e}_{P'}^*|\Psi_P(t')\rangle \mathrm{e}^{\mathrm{i}\phi_P} \tag{4.14a}$$

$$= 2\Im\left\{\underbrace{\mathrm{i}\int_{-\infty}^{\infty} \mathrm{d}t' \langle \Psi_0(t')|(-\mathrm{i})\varepsilon_{P'}^*(t'-t_{P'})\boldsymbol{\mu}\cdot\mathbf{e}_{P'}^*\mathrm{e}^{\mathrm{i}H_0(t-t')}}_{\approx\langle\Psi_{P'}(t)|}\right.$$

$$\left.\times\underbrace{\mathrm{e}^{-\mathrm{i}H_0(t-t')}|\Psi_P(t')\rangle}_{=|\Psi_P(t)\rangle}\mathrm{e}^{\mathrm{i}(\phi_P-\phi_{P'})}\right\} \tag{4.14b}$$

$$\approx 2\Re\langle\Psi_{P'}|\Psi_P\rangle \mathrm{e}^{\mathrm{i}(\phi_P-\phi_{P'})}, \tag{4.14c}$$

where in equation (4.14*b*) we have approximated $|\Psi_{P'}(t)\rangle$ by taking the integral $\int_{-\infty}^{t}\mathrm{d}t' \approx \int_{-\infty}^{\infty}\mathrm{d}t'$, choosing t such that $t - t_{P'} \gg \sigma$, and inserting the identity in the form $\mathbb{I} = \mathrm{e}^{-\mathrm{i}H_0(t-t')}\mathrm{e}^{\mathrm{i}H_0(t-t')}$. This approximation is possible only because $(t_{P'} - t_P)/\sigma \gg 1$, the pulses are well separated (assumption from equation 3.10). In equation (4.14*c*) we have used equation (4.3). The second term in equation (4.12) is easily seen to be just $\langle\Psi_{P'}|\Psi_{P'}\rangle$. Altogether,

$$S_{P'} = 2\Re\langle\Psi_{P'}|\Psi_P\rangle \mathrm{e}^{\mathrm{i}(\phi_P-\phi_{P'})} + \langle\Psi_{P'}|\Psi_{P'}\rangle. \tag{4.15}$$

3. The difference between the answers in parts 1 and 2 can be traced back to phase-matching. Let us examine this a bit more carefully. The excited-state amplitude, to first order in η, is,

$$|\Psi^{(1)}(\mathbf{r},t)\rangle = \mathrm{e}^{\mathrm{i}\mathbf{k}_P\cdot\mathbf{r}+\mathrm{i}\phi_P}|\Psi_P(t)\rangle + \mathrm{e}^{\mathrm{i}\mathbf{k}_{P'}\cdot\mathbf{r}+\mathrm{i}\phi_{P'}}|\Psi_{P'}(t)\rangle \tag{4.16}$$

so that the population in the SEM is,

$$\langle\Psi^{(1)}(\mathbf{r},t)|\Psi^{(1)}(\mathbf{r},t)\rangle = \underbrace{\langle\Psi_P|\Psi_P\rangle}_{=N_P(\mathbf{r})} + \underbrace{\langle\Psi_{P'}|\Psi_{P'}\rangle + 2\Re \mathrm{e}^{\mathrm{i}(\mathbf{k}_P-\mathbf{k}_{P'})\cdot\mathbf{r}+\mathrm{i}(\phi_P-\phi_{P'})}\langle\Psi_{P'}|\Psi_P\rangle}_{=N_{P'}(\mathbf{r})}, \tag{4.17}$$

where $N_P(\mathbf{r})$ and $N_{P'}(\mathbf{r})$ are the number of photons of pulses P and P' that are absorbed by the molecule at location $\mathbf{r}$ (equation (3.48)), and only $N_{P'}(\mathbf{r})$ is a function of $\mathbf{r}$ in this particular example. Clearly, when $\mathbf{k}_P \neq \mathbf{k}_{P'}$ (noncollinear setting), the interference term will vary as a function of $\mathbf{r}$, so molecules at different positions will absorb different amounts of energy. However, what matters in the experimental measurement is the average absorption throughout the ensemble. The interference term will average to zero and only the spatially independent terms survive, as expected from our previous discussion (equation (3.50)). When $\mathbf{k}_P = \mathbf{k}_{P'}$ (collinear setting), the interference term does not depend on spatial location and therefore also survives the averaging. Furthermore, the average amount of absorbed light across the sample will change as a function of the waiting time T due to the time dependence of $\langle \Psi_{P'} | \Psi_P \rangle$. This comparison of the collinear and noncollinear setups yields an intuitive picture of phase-matching.

4.2 Pump–probe (*PP′*) spectroscopy

The next situation we consider is a PP' experiment like the one illustrated in figure 4.4. Here, a *pump* pulse P traveling in the $\mathbf{k}_P$ direction excites the molecules from their ground to their bright excited states via the dipole operator. Nonequilibrium dynamics ensues, such as the activation of certain vibrational modes of the bath and the transfer of amplitude to other bright and dark states. How do we detect these processes? After a waiting time T, a second short *probe* pulse P' traveling in the $\mathbf{k}_{P'}$ direction passes through the sample, and the number of photons gained or lost is measured in a photodetector. We are interested in a differential signal $S_{PP'}$, namely, the number of photons lost by the probe P' after the pump P has acted ($S_{P'}$ (with P)), minus the ones lost by P' if P is not present ($S_{P'}$ (without P)). Hence, this technique is known by a variety of names, including transient, differential, or dynamic absorption. The PP' signal is given, to lowest nonvanishing order in η, by [3, 4, 9, 10, 13]

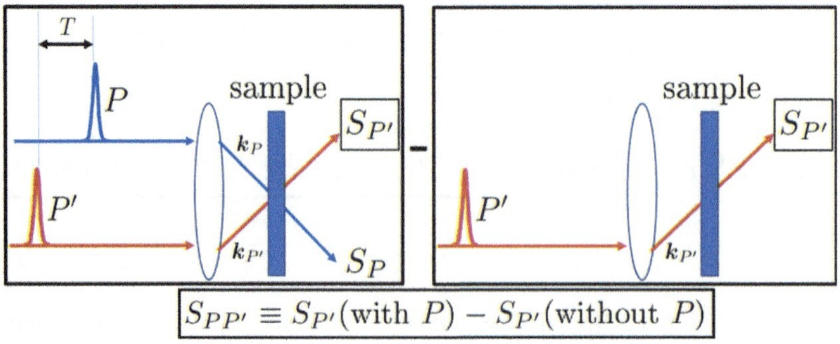

Figure 4.4. Pump–probe (PP') spectroscopy. A pump P excites the sample. After a waiting time T, the absorption of a probe P' is measured. To obtain a differential measurement, the background absorption of P' in the absence of P is subtracted.

$$\begin{aligned} S_{PP'} &\equiv S_{P'}(\text{with}\, P) - S_{P'}(\text{without}\, P) \\ &= 2\Im \int_{-\infty}^{\infty} dt' \varepsilon_{P'}^{*}(t' - t_{P'}) e^{-i\phi_{P'}} \mathbf{e}_{P'}^{*} \cdot [\mathbf{P}_{\mathbf{k}_{P'}}(t') - \mathbf{P}_{\mathbf{k}_{P'}}^{(1)}] e^{i\phi_{P'}} \\ &= 2\Im \int_{-\infty}^{\infty} dt' \varepsilon_{P'}^{*}(t' - t_{P'}) \mathbf{e}_{P'}^{*} \cdot \mathbf{P}_{\mathbf{k}_{P}}^{(3)}(t'). \end{aligned} \tag{4.18}$$

We have noticed that $S_{P'}$ (without P) is simply the linear absorption of P', which means that the lowest order contribution of $\mathbf{P}_{\mathbf{k}_P}$ to the differential absorption $S_{PP'}$ will be $\mathbf{P}_{\mathbf{k}_P}^{(3)}$, the $O(\eta^3)$ polarization along the direction of the probe. Previously, we calculated this nonlinear polarization for the model of vibrationless coupled dimers (see section 3.2). The result from equation (3.33) holds more generally than the context of that section,

$$\begin{aligned} P_{\mathbf{k}_{P'}}^{(3)}(t) = {} & \langle \Psi_{+P}(t') | \boldsymbol{\mu} | \Psi_{+P+P'}(t') \rangle + \langle \Psi_{+P-P'}(t') | \boldsymbol{\mu} | \Psi_{+P}(t') \rangle \\ & + \langle \Psi_{+P-P}(t') | \boldsymbol{\mu} | \Psi_{+P'}(t) \rangle + \langle \Psi_{0}(t) | \boldsymbol{\mu} | \Psi_{+P-P+P'}(t') \rangle, \end{aligned} \tag{4.19}$$

as well as the general definitions for the various wavefunctions in its calculation (equations (3.18*a*)–(3.31*a*)), with the exception that, this time, each wavefunction is in the $\mathscr{S}$ and $\mathscr{B}$ space, instead of only in the $\mathscr{S}$ space. In addition, we note from equation (4.18) that $S_{PP'}$ does not depend on the phases ϕ_n of the different pulses.

We emphasize again that this is a frequency-integrated result, where we are uniformly counting all the photons lost by P' due to material absorption, regardless of their color. We refer the reader to appendix D and specifically to example 14 to learn about the *frequency-resolved* version of PP' spectroscopy.

Example 5. PP' signal in terms of wavepacket overlaps

By physically interpreting the processes that contribute to the PP' signal, we may express equation (4.18) in a more intuitively appealing form. First, we define the projectors onto the respective excitation manifolds,

$$\begin{aligned} \mathbb{P}_{\text{GSM}} &= \sum_{m \in \text{GSM}} |m\rangle\langle m|, \\ \mathbb{P}_{\text{DEM}} &= \sum_{m \in \text{DEM}} |m\rangle\langle m|, \end{aligned}$$

as well as the wavefunctions,

$$|\Psi_{P'}(t)\rangle = i \int_{-\infty}^{t} dt' e^{-iH_0(t-t')} \left[\boldsymbol{\mu} \cdot \mathbf{e}_{P'}(\varepsilon_{P'}(t' - t_{P'}) + \text{c.c.})\right] |\Psi_0(t')\rangle, \tag{4.20}$$

$$|\Psi_{Pn}(t)\rangle = i \int_{-\infty}^{t} dt' e^{-iH_0(t-t')} \left[\boldsymbol{\mu} \cdot \mathbf{e}_{n}(\varepsilon_{n}(t' - t_{n}) + \text{c.c.})\right] |\Psi_P(t')\rangle \quad \text{for } n = P, P', \tag{4.21}$$

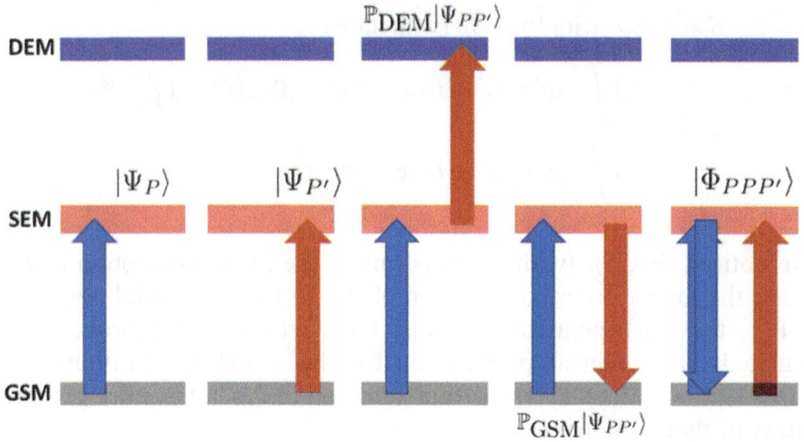

Figure 4.5. Wavepackets involved in the calculation of the frequency integrated PP' signal. The blue and the red arrows denote actions of the pump P and the probe P' respectively.

$$|\Phi_{PPP'}(t)\rangle = \mathrm{i}\int_{-\infty}^{t} \mathrm{d}t' \mathrm{e}^{-\mathrm{i}H_0(t-t')}[\boldsymbol{\mu}\cdot\mathbf{e}_{P'}(\varepsilon_{P'}(t'-t_{P'})+\mathrm{c.c.})]\mathbb{P}_{\mathrm{GSM}}|\Psi_{PP}(t')\rangle. \tag{4.22}$$

Here, $|\Psi_{P'}(t)\rangle$ and $|\Psi_{Pn}(t)\rangle$ are analogues of $|\Psi_P(t)\rangle$ (equation (4.2)). In the RWA, $|\Psi_n(t)\rangle$ is a wavepacket in the SEM created by the action of pulse n alone, and $|\Psi_{Pn}(t)\rangle$ is a wavepacket in a superposition of the GSM and the DEM, created by both P and n. $|\Phi_{PPP'}(t)\rangle$ is a wavepacket that results from the action of P' on $\mathbb{P}_{\mathrm{GSM}}|\Psi_{PP}(t')\rangle$, which is the projection of $|\Psi_{PP}(t')\rangle$ onto the ground state; therefore, it corresponds to a wavepacket in the SEM. Importantly, $|\Phi_{PPP'}(t)\rangle$ differs from what would be called $|\Psi_{PPP'}(t)\rangle$, since the latter consists also of additional contributions due to the action of P' on $\mathbb{P}_{\mathrm{DEM}}|\Psi_{PP}(t')\rangle$, and they do not satisfy the phase-matching requirements of the PP' signal.

Also, in analogy to the definition of $|\Psi_P\rangle$ in equation (4.3), we define the asymptotic wavepackets,

$$|\Psi_{P'}\rangle \equiv \lim_{t\to\infty} \mathrm{e}^{\mathrm{i}H_0 t}|\Psi_{P'}(t)\rangle, \tag{4.23}$$

$$|\Psi_{Pn}\rangle \equiv \lim_{t\to\infty} \mathrm{e}^{\mathrm{i}H_0 t}|\Psi_{Pn}(t)\rangle \quad \text{for } n = P, P', \tag{4.24}$$

$$|\Phi_{PPP'}\rangle \equiv \lim_{t\to\infty} \mathrm{e}^{\mathrm{i}H_0 t}|\Phi_{PPP'}(t)\rangle. \tag{4.25}$$

See figure 4.5 for diagrams for each of these wavefunctions. Our goal is to show that the PP' signal $S_{PP'}$ may be rewritten as a sum of terms: excited-state absorption (ESA), stimulated emission (SE) and ground-state bleach (GSB),

$$S_{PP'} = S_{\mathrm{SE}} + S_{\mathrm{ESA}} + S_{\mathrm{GSB}}, \tag{4.26}$$

where each of the terms is given by the wavefunction overlaps,

$$S_{\mathrm{ESA}} \equiv 2\Im\int_{-\infty}^{\infty} \mathrm{d}t' \varepsilon_{P'}^{*}(t'-t_{P'})\langle\Psi_{+P}(t')|\boldsymbol{\mu}\cdot\mathbf{e}_{P'}^{*}|\Psi_{+P+P'}(t')\rangle \tag{4.27a}$$

$$= \langle\Psi_{PP'}|\mathbb{P}_{\mathrm{DEM}}|\Psi_{PP'}\rangle, \tag{4.27b}$$

$$S_{\mathrm{SE}} \equiv 2\Im\int_{-\infty}^{\infty} \mathrm{d}t' \varepsilon_{P'}^{*}(t'-t_{P'})\langle\Psi_{+P-P'}(t')|\boldsymbol{\mu}\cdot\mathbf{e}_{P'}^{*}|\Psi_{+P}(t')\rangle \tag{4.28a}$$

$$= -\langle\Psi_{PP'}|\mathbb{P}_{\mathrm{GSM}}|\Psi_{PP'}\rangle, \tag{4.28b}$$

$$S_{\mathrm{GSB}} \equiv 2\Im\int_{-\infty}^{\infty} \mathrm{d}t' \varepsilon_{P'}^{*}(t'-t_{P'})\big[\langle\Psi_{+P-P}(t')|\boldsymbol{\mu}\cdot\mathbf{e}_{P'}^{*}|\Psi_{+P'}(t)\rangle$$

$$+ \langle\Psi_{0}(t)|\boldsymbol{\mu}\cdot\mathbf{e}_{P'}^{*}|\Psi_{+P-P+P'}(t')\rangle\big] \tag{4.29a}$$

$$= 2\Re\langle\Phi_{PPP'}|\Psi_{P'}\rangle. \tag{4.29b}$$

Important: The assumptions from section 3.2, equations (3.9) and (3.10), still hold, but equation (3.11) *does not*. The reason is that the window of time where the integrand of equation (4.18) is appreciable is determined by $\varepsilon_{P'}^{*}(t'-t_{P'})$, that is, for $t' \in [t_{P'} - 3\sigma_{P'}, t_{P'} + 3\sigma_{P'}]$, which is *not* what equation (3.11) conveys. Hence, do not assume $\int_{-\infty}^{t'} \approx \int_{-\infty}^{\infty}$ when evaluating integrals with respect to $\varepsilon_{P'}^{*}$ in $\mathbf{P}_{\mathbf{k}_P}^{(3)}(t')$. *Hint:* consult the derivation of equation (4.5).

Qualitatively discuss the type of dynamics that each of the contributions to $S_{PP'}(T)$ corresponds to.

Solution

In the RWA, the following identities hold,

$$|\Psi_{n}(t)\rangle \approx |\Psi_{+n}(t)\rangle, \tag{4.30}$$

$$\mathbb{P}_{\mathrm{GSM}}|\Psi_{Pn}(t)\rangle \approx |\Psi_{+P-n}(t)\rangle, \tag{4.31}$$

$$\mathbb{P}_{\mathrm{DEM}}|\Psi_{Pn}(t)\rangle \approx |\Psi_{+P+n}(t)\rangle, \tag{4.32}$$

$$|\Phi_{PPP'}(t)\rangle \approx |\Psi_{+P-P+P'}(t')\rangle. \tag{4.33}$$

For instance, $|\Psi_{PP}(t)\rangle$ in principle consists of all the wavefunctions $|\Psi_{\pm P\pm P}(t)\rangle$. Since the initial state $|\Psi_0(t)\rangle$ is in the GSM, it can only interact, in the RWA, with the resonant term ε_P rather than ε_P^*, gaining energy to transition to the SEM (equation (3.25)); hence, we can ignore the wavefunctions associated with a first interaction with ε_P^*, $|\Psi_{-P\pm P}(t)\rangle \approx 0$. Similarly, once in the SEM, the state of the molecule can gain energy via $\varepsilon_{P'}$ to transition to the DEM, or lose energy to the field via $\varepsilon_{P'}^*$ and transition back to the GSM, yielding $|\Psi_{P+P}(t)\rangle$ or $|\Psi_{P-P}(t)\rangle$, respectively. The rest of the identities in equations (4.30)–(4.33) can be similarly understood.

Upon substitution of equations (4.30)–(4.33) into equation (3.33) or (4.19) and subsequently into equation (4.18), we get a sum of contributions that we can match to the structure of equation (4.26) via the identifications in equations (4.27*a*)–(4.29*b*),

$$
\begin{aligned}
S_{\rm ESA} &= 2\Im\int_{-\infty}^{\infty} dt'\varepsilon^*_{P'}(t'-t_{P'})\langle\Psi_{+P}(t')|\boldsymbol{\mu}\cdot\mathbf{e}^*_{P'}|\Psi_{+P+P'}(t')\rangle \\
&= 2\Im\left\{ i\int_{-\infty}^{\infty} dt'\langle\Psi_P(t')|(-i)\varepsilon^*_{P'}(t'-t_{P'})\boldsymbol{\mu}\cdot\mathbf{e}^*_{P'}|\mathbb{P}_{\rm DEM}|\Psi_{PP'}(t')\rangle\right\} \\
&= 2\Im\left\{ i\int_{-\infty}^{\infty} dt'[\partial_{t'}\langle\Psi_{PP'}(t')|]\mathbb{P}_{\rm DEM}|\Psi_{PP'}(t')\rangle\right\} \\
&= \langle\Psi_{PP'}|\mathbb{P}_{\rm DEM}|\Psi_{PP'}\rangle,
\end{aligned} \tag{4.34}
$$

$$
\begin{aligned}
S_{\rm SE} &= 2\Im\int_{-\infty}^{\infty} dt'\varepsilon^*_{P'}(t'-t_{P'})\langle\Psi_{+P-P'}(t')|\boldsymbol{\mu}\cdot\mathbf{e}^*_{P'}|\Psi_{+P}(t')\rangle \\
&= 2\Im\left\{(-i)\int_{-\infty}^{\infty} dt'\langle\Psi_{PP'}(t')|\mathbb{P}_{\rm GSM} i\varepsilon^*_{P'}(t'-t_{P'})\boldsymbol{\mu}\cdot\mathbf{e}^*_{P'}|\Psi_P(t')\rangle\right\} \\
&= 2\Im\left\{(-i)\int_{-\infty}^{\infty} dt'\langle\Psi_{PP'}(t')|\mathbb{P}_{\rm GSM}[\partial_{t'}|\Psi_{PP'}(t')\rangle]\right\} \\
&= -\langle\Psi_{PP'}|\mathbb{P}_{\rm GSM}|\Psi_{PP'}\rangle,
\end{aligned} \tag{4.35}
$$

$$
\begin{aligned}
S_{\rm GSB} &= 2\Im\int_{-\infty}^{\infty} dt'\varepsilon^*_{P'}(t'-t_{P'}) \\
&\quad\times[\langle\Psi_{+P-P}(t')|\boldsymbol{\mu}\cdot\mathbf{e}^*_{P'}|\Psi_{+P'}(t)\rangle + \langle\Psi_0(t)|\boldsymbol{\mu}\cdot\mathbf{e}^*_{P'}|\Psi_{+P-P+P'}(t')\rangle] \\
&= 2\Im\left\{ i\int_{-\infty}^{\infty} dt'\langle\Psi_{PP}(t')|\mathbb{P}_{\rm GSM}(-i)\varepsilon^*_{P'}(t'-t_{P'})\boldsymbol{\mu}\cdot\mathbf{e}^*_{P'}|\Psi_{P'}(t')\rangle\right\} \\
&\quad + 2\Im\left\{ i\int_{-\infty}^{\infty} dt'\langle\Psi_0(t')|(-i)\varepsilon^*_{P'}(t'-t_{P'})\boldsymbol{\mu}\cdot\mathbf{e}^*_{P'}|\Phi_{PPP'}(t')\rangle\right\} \\
&= 2\Re\int_{-\infty}^{\infty} dt'[\partial_{t'}\langle\Phi_{PPP'}(t')|]|\Psi_{P'}(t')\rangle + [\partial_{t'}\langle\Psi_{P'}(t')|]|\Phi_{PPP'}(t')\rangle \\
&= 2\Re\langle\Phi_{PPP'}|\Psi_{P'}\rangle,
\end{aligned} \tag{4.36}
$$

confirming equations (4.27*b*)–(4.29*b*). See figure 4.5 for a schematic view of this result via DS-FDs. Here we have employed similar techniques for the derivation of equation (4.5).

The interpretation of these equations is as follows. P promotes amplitude from $|g\rangle$ to $|\Psi_P(t)\rangle$, a wavepacket consisting of amplitude in the SEM. P' acts on this state,

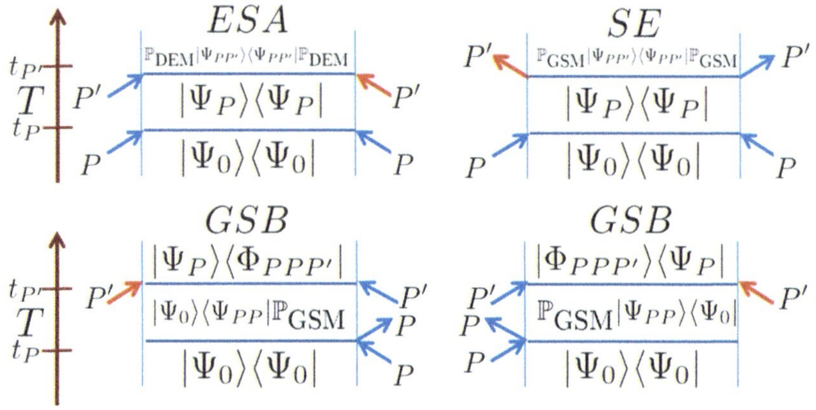

Figure 4.6. DS-FDs for the different contributions of the PP' signal.

creating $|\Psi_{PP'}(t)\rangle$, a superposition of wavepackets in the GSM and DEM. Naturally, the number of photons *absorbed* by the material through ESA corresponds to the population transferred to the DEM, that is, the square amplitude of $\mathbb{P}_{\mathrm{DEM}}|\Psi_{PP'}(t)\rangle$. Analogously, the squared amplitude of $\mathbb{P}_{\mathrm{GSM}}|\Psi_{PP'}(t)\rangle$ corresponds to the photons *gained* by P' via SE. Hence, $S_{\mathrm{ESA}}(T)$ and $S_{\mathrm{SE}}(T)$ are always positive and negative contributions to $S_{PP'}(T)$, respectively, since $S_{PP'}$ is defined as the photons lost by P'. $S_{\mathrm{GSB}}(T)$ seems a bit trickier to interpret but it can be analyzed similarly to the other processes. Consider the wavepacket left in the ground state after P has acted but before P' has. It reads:

$$\mathbb{P}_{\mathrm{GSM}}|\Psi(t)\rangle = \mathbb{P}_{\mathrm{GSM}}(|\Psi_0(t)\rangle + |\Psi_{PP}(t)\rangle), \tag{4.37}$$

where we have used the fact that the only wavepackets that have amplitude in the GSM are $|\Psi_0(t)\rangle$[3] and $|\Psi_{PP}(t)\rangle$. This wavepacket can absorb energy from P' to transition to the SEM, creating the wavefunction,

$$\mathrm{i}\int_{-\infty}^{\infty} \mathrm{d}t'[\varepsilon^*_{P'}(t'-t_{P'})\boldsymbol{\mu}\cdot\mathbf{e}^*_{P'}]\mathbb{P}_{\mathrm{GSM}}|\Psi(t')\rangle = |\Psi_{P'}(t)\rangle + |\Phi_{PPP'}(t)\rangle. \tag{4.38}$$

The number of photons absorbed through this process is equal to the population created in the SEM once P' has been fully absorbed,

$$\begin{aligned}
&\text{Number of photons of } P' \text{ absorbed to go from GSM to SEM}\\
&\quad= \lim_{t\to\infty} \mathrm{e}^{\mathrm{i}H_0 t}\big((\langle\Psi_{P'}(t)| + \langle\Phi_{PPP'}(t)|)(|\Psi_{P'}(t)\rangle + |\Phi_{PPP'}(t)\rangle)\big)\\
&\quad= \langle\Psi_{P'}|\Psi_{P'}\rangle + 2\Re\{\langle\Phi_{PPP'}|\Psi_{P'}\rangle\} + O(\eta^6)\\
&\quad\approx S_{P'}(\text{without } P) + S_{\mathrm{GSB}}
\end{aligned} \tag{4.39}$$

[3] Note that $|\Psi_0(t)\rangle = \mathbb{P}_{\mathrm{GSM}}|\Psi_0(t)\rangle$.

where we have adapted equation (4.5) for the linear absorption of P', $S_{P'}$ (without P), and used equation (4.29*b*) for S_{GSB}. Also, we have ignored $\langle\Phi_{PPP'}|\Phi_{PPP'}\rangle$ as an $O(\eta^6)$ contribution. Hence, the total absorption of P' is given by

$$S_{P'}(\text{with } P) \approx S_{\text{SE}} + S_{\text{ESA}} + S_{P'}(\text{without } P) + S_{\text{GSB}}. \tag{4.40}$$

However, by definition (equation (4.18)), $S_{PP'}$ denotes only the differential absorption. Subtracting the linear absorption $S_{P'}$ (without P) from equation (4.40), finally yields equation (4.26). Even though equation (4.39), being the norm squared of a wavepacket, is always positive, S_{GSB} does not need to be positive due to the possibility of interferences in the overlap $\langle\Phi_{PPP'}|\Psi_{P'}\rangle$.

The contributions due to equations (4.27*b*)–(4.29*b*) clearly depend on the waiting time T, which may be sampled by repeating the experiment multiple times. Since $|\Psi_P(t)\rangle$ describes a wavefunction in the SEM, $|\Psi_{PP'}\rangle$ (and consequently $S_{\text{SE}}(T)$ and $S_{\text{ESA}}(T)$) directly report on dynamics in the SEM as a function of T. Qualitatively, P launches amplitude from the GSM to the SEM, and coherent dynamics in the SEM follows throughout the waiting time T. Via P', we probe these amplitudes by transferring them into the DEM (S_{ESA}) or back into the GSM (S_{SE}). The number of photons of P' that get absorbed is equal to the population created in the DEM minus that which ended in the GSM. These populations depend on the interferences between the different transferred amplitudes, and hence, directly report on the dynamics in the SEM.

Let us now turn to $S_{\text{GSB}}(T)$. Both $\mathbb{P}_{\text{GSM}}|\Psi_0\rangle$ and $\mathbb{P}_{\text{GSM}}|\Psi_{PP}(t)\rangle$ refer to wavefunctions in the GSM. Even though $\mathbb{P}_{\text{GSM}}|\Psi_0\rangle$ is by definition a stationary wavefunction, $\mathbb{P}_{\text{GSM}}|\Psi_{PP}(t)\rangle$ need not be: if P is *not* ideally broadband (i.e., P is not a delta function in time), the two actions of P on $|\Psi_0\rangle$ may distort the shape of the wavepacket and leave it in a superposition of eigenstates of H_g. After time T, P' brings both of these wavefunctions to the SEM, creating $|\Psi_{P'}(t)\rangle$ and $|\Phi_{PPP'}(t)\rangle$, respectively, and $S_{\text{GSB}}(T)$ depends on $\langle\Phi_{PPP'}|\Psi_{P'}\rangle$. If P' is broadband, $\varepsilon_{P'}(t - t_{P'}) \propto \delta(t - t_{P'})$, then $\langle\Phi_{PPP'}|\Psi_{P'}\rangle \propto \langle\Psi_{PP}(t)|\Psi_0(t)\rangle$ and the GSB signal directly tests the nonstationary dynamics of $|\Psi_{PP}(t)\rangle$ projected onto the reference wavefunction $|\Psi_0(t)\rangle$. If P' is not ideally broadband, the correlation function $\langle\Psi_{PP}(t)|\Psi_0(t)\rangle$ will be reweighted in frequency domain by the corresponding transition amplitudes in $\langle\Phi_{PPP'}|\Psi_{P'}\rangle$. Either way, $S_{\text{GSB}}(T)$ is a reporter on GSM dynamics as a function of T.[4]

Although $S_{PP'}(T)$ can be expressed in general as a sum of ESA, SE and GSB contributions (equation (4.26)), if the transition energies between the different excitation manifolds are sufficiently different, one can imagine isolating a particular contribution by controlling the frequency components of the pulses. The reader is encouraged to explore these possibilities.

[4] Throughout this book, we are not considering processes that transfer amplitude from one excitation manifold to another in the absence of optical pulses within the timescales of the experiment. However, it is possible that, say, after the pump P transfers amplitude from the GSM to the SEM, an ultrafast nonradiative pathway from the SEM replenishes the GSM during the waiting time T. If decoherence in the GSM sets in fast enough so that we can ignore the interferences between the various nuclear wavepackets in it, the absorption of the probe P' will be *increased* (compared to when this pathway is off) due to the recovered population in the GSM. This process is naturally called ground-state recovery (GSR), and can be regarded as opposite to GSB. We shall ignore GSR processes for the rest of this book.

Example 6. *PP′* signal for an vibrationless coupled dimer

Evaluate equations (4.27*b*)–(4.29*b*) for the vibrationless coupled dimer (where the electronic system $\mathscr{S}$ is not coupled to vibrations $\mathscr{B}$) from section 3.2 and discuss the results.

Solution

Adapting the wavefunctions for a vibrationless coupled dimer from section 3.2 (equations (3.18*a*)–(3.31*a*)), together with the identifications in equations (4.31)–(4.33) and the asymptotic wavepackets in equations (4.23)–(4.25), yields[5]

$$|\Psi_{P'}\rangle = \mathrm{i}\sum_{q=\alpha,\beta} \mathrm{e}^{\mathrm{i}\omega_q t_{P'}}|q\rangle\Omega^{P'}_{qg}\mathrm{e}^{-\mathrm{i}\omega_g t_{P'}}, \tag{4.41}$$

$$\begin{aligned}|\Psi_{Pn}\rangle = &-\mathrm{e}^{\mathrm{i}\omega_g t_n}|g\rangle\sum_{q=\alpha,\beta}\Omega^{\bar{n}}_{gq}\mathrm{e}^{-\mathrm{i}\omega_q(t_n-t_P)}\Omega^{P}_{qg}\mathrm{e}^{-\mathrm{i}\omega_g t_P}\\ &-\mathrm{e}^{\mathrm{i}\omega_f t_n}|f\rangle\sum_{q=\alpha,\beta}\Omega^{n}_{fq}\mathrm{e}^{-\mathrm{i}\omega_q(t_n-t_P)}\Omega^{P}_{qg}\mathrm{e}^{-\mathrm{i}\omega_g t_P}, \quad \text{for } n=P,P',\end{aligned} \tag{4.42}$$

$$\begin{aligned}|\Phi_{PPP'}\rangle = &-\mathrm{i}\sum_{i=\alpha,\beta}\mathrm{e}^{\mathrm{i}\omega_i t_{P'}}|i\rangle\Omega^{P'}_{ig}\mathrm{e}^{-\mathrm{i}\omega_g(t_{P'}-t_P)}\\ &\times\left(\sum_{q=\alpha,\beta}\frac{\Omega^{\bar{P}}_{gq}\Omega^{P}_{qg}}{2}\left[1-\mathrm{erf}(\mathrm{i}(\omega_{qg}-\omega_P)\sigma_P)\right]\right)\mathrm{e}^{-\mathrm{i}\omega_g t_P}.\end{aligned} \tag{4.43}$$

Using the facts that in the ideal coupled dimer model, $\mathbb{P}_{\mathrm{GSM}} = |g\rangle\langle g|$ and $\mathbb{P}_{\mathrm{DEM}} = |f\rangle\langle f|$, we can readily substitute equations (4.41)–(4.43) into equations (4.27*b*)–(4.29*b*) to obtain (see DS-FDs in figure 4.7),

$$S_{\mathrm{ESA}}(T) = \sum_{p,q=\alpha,\beta}\Omega^{P'}_{fq}\Omega^{P}_{qg}\Omega^{\bar{P}}_{gp}\Omega^{\bar{P}'}_{pf}\mathrm{e}^{-\mathrm{i}\omega_{qp}T}, \tag{4.44a}$$

$$S_{\mathrm{SE}}(T) = -\sum_{p,q=\alpha,\beta}\Omega^{\bar{P}'}_{gq}\Omega^{P}_{qg}\Omega^{\bar{P}}_{gp}\Omega^{\bar{P}'}_{pg}\mathrm{e}^{-\mathrm{i}\omega_{qp}T}, \tag{4.44b}$$

$$S_{\mathrm{GSB}}(T) = -\sum_{p,q=\alpha,\beta}\Omega^{P'}_{qg}\Omega^{\bar{P}}_{gp}\Omega^{P}_{pg}\Omega^{\bar{P}'}_{gq}. \tag{4.44c}$$

In deriving equation (4.44*c*), we used the fact that $\mathrm{erf}(\mathrm{i}x)$ is purely imaginary for a real valued x.

The results above may also be obtained by directly evaluating each of the contributions of $S_{PP'}(T)$ using equations (4.27*a*), (4.28*a*) and (4.29). The relevant

[5] We recall that the asymptotic wavepackets from equations (4.23)–(4.25) include a phase shift via $\mathrm{e}^{\mathrm{i}H_0 t}$. For instance, even though the final state in equation (4.41) is a superposition of $|q\rangle$ states, and each of them evolves from time $t_{P'}$ up to t as $\mathrm{e}^{-\mathrm{i}\omega_q(t-t_{P'})}|q\rangle$, we need to subsequently translate them backwards in time for a period of time t, $\lim_{t\to\infty}\mathrm{e}^{\mathrm{i}H_0 t}(\mathrm{e}^{-\mathrm{i}\omega_q(t-t_{P'})}|q\rangle) = \lim_{t\to\infty}\mathrm{e}^{\mathrm{i}\omega_q t}\mathrm{e}^{-\mathrm{i}\omega_q(t-t_{P'})}|q\rangle = \mathrm{e}^{\mathrm{i}\omega_q t_{P'}}|q\rangle$, so that the indefinite oscillatory dependence on t gets eliminated when taking the corresponding limit.

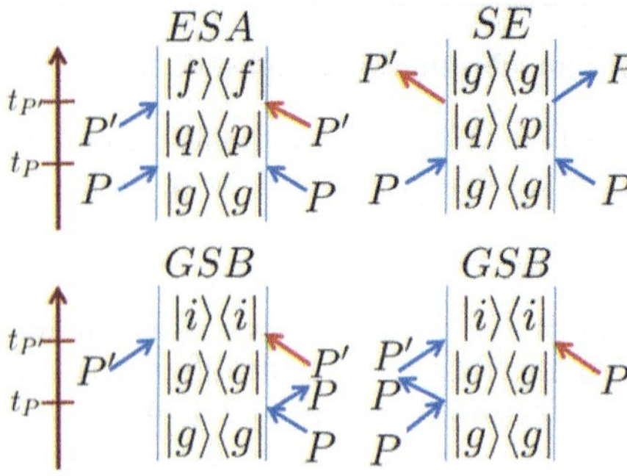

Figure 4.7. DS-FDs for ideal coupled dimer PP' spectra. Note that coherences or populations do not decay in this model, as coupling to the bath is not present in this model.

wavefunctions are given by the general definitions in equations (3.18*a*), (3.19*a*), (3.28*a*), (3.29*a*), (3.30*a*) and (3.31*a*). As we have discussed, we cannot exploit the final results from section 3.2 in their entirety, as they assume equation (3.11), which does not hold in the evaluation of $S_{PP'}$ (see note under **Important** in example 5).

As an example, let us explicitly compute $S_{\text{ESA}}(T)$ from equation (4.27*a*). In order to do so, we need expressions for $|\Psi_{+P}(t)\rangle$ and $|\Psi_{+P+P'}(t)\rangle$. For $|\Psi_{+P}(t)\rangle$ we *can* actually use the final result of equation (3.19*c*) (setting $n = P$), as pulse P is well separated from pulse P' (assumption from equation (3.10)), and it does not depend on equation (3.11) being true (an integral over $\varepsilon_{P'}$ is not involved). However, we cannot directly use equation (3.29*b*) for $|\Psi_{+P+P'}(t)\rangle$. Instead,

$$\begin{aligned}|\Psi_{+P+P'}(t)\rangle &= (-\mathrm{i})^2\int_0^t \mathrm{d}t'\int_0^{t'}\mathrm{d}t''\mathrm{e}^{-\mathrm{i}H_0(t-t')}[-\boldsymbol{\mu}\cdot\mathbf{e}_{P'}\varepsilon_{P'}(t'-t_{P'})]\\ &\quad\times\mathrm{e}^{-\mathrm{i}H_0(t'-t'')}[-\boldsymbol{\mu}\cdot\mathbf{e}_P\varepsilon_P(t''-t_P)]\mathrm{e}^{-\mathrm{i}H_0t''}|\Psi_0\rangle\\ &= -|f\rangle\sum_{q=\alpha,\beta}\mathrm{e}^{-\mathrm{i}\omega_f(t-t_{P'})}\int_{-\infty}^{t}\mathrm{d}t''\mathrm{e}^{\mathrm{i}\omega_{fq}(t''-t_P)}\varepsilon_{P'}(t''-t_{P'})\\ &\quad\times(\boldsymbol{\mu}_{fq}\cdot\mathbf{e}_{P'})\mathrm{e}^{-\mathrm{i}\omega_qT}\Omega^P_{qg}\mathrm{e}^{-\mathrm{i}\omega_gt_P},\end{aligned}\tag{4.45}$$

where the integral over t'' has been approximated by extending its limits to go from $-\infty$ to ∞, but the one over t' only goes from $-\infty$ to t, since it is associated with the action of $\varepsilon_{P'}$. Therefore, the relevant matrix element for S_{ESA} is

$$\begin{aligned}\langle\Psi_{+P}(t')|\boldsymbol{\mu}\cdot\mathbf{e}^*_{P'}|\Psi_{+P+P'}(t')\rangle &= \mathrm{i}\sum_{p,q=\alpha,\beta}\int_{-\infty}^{t}\mathrm{d}t''\mathrm{e}^{\mathrm{i}\omega_{fq}(t''-t_P)}\varepsilon_{P'}(t''-t_{P'})\\ &\quad\times(\boldsymbol{\mu}_{fq}\cdot\mathbf{e}_{P'})\Omega^P_{qg}\Omega^{\overline{P}}_{gp}(\boldsymbol{\mu}_{pf}\cdot\mathbf{e}^*_{P'})\mathrm{e}^{-\mathrm{i}\omega_{fp}(t'-t_{P'})}\mathrm{e}^{-\mathrm{i}\omega_{qp}T}.\end{aligned}\tag{4.46}$$

By exploiting the nested integral (see footnote 5 on page 3–10),

$$\begin{aligned}&\int_{-\infty}^{\infty}\mathrm{d}t'\varepsilon^*_{P'}(t'-t_{P'})\mathrm{e}^{-\mathrm{i}\omega_{fp}(t'-t_{P'})}\int_{-\infty}^{t'}\mathrm{d}t''\varepsilon_{P'}(t''-t_{P'})\mathrm{e}^{\mathrm{i}\omega_{fq}(t''-t_{P'})}\\ &= \frac{1}{2}\tilde{\varepsilon}_{P'}(\omega_{fq})\tilde{\varepsilon}^*_{P'}(\omega_{fp})\left\{1-\mathrm{erf}\left(\frac{\mathrm{i}(\omega_{fp}+\omega_{fq}-2\omega_{P'})\sigma_{P'}}{2}\right)\right\},\end{aligned}\tag{4.47}$$

we finally obtain,

$$S_{\text{ESA}} = 2\Im \int_{-\infty}^{\infty} \mathrm{d}t' \varepsilon_{P'}^{*}(t' - t_{P'}) \langle \Psi_{+P}(t') | \boldsymbol{\mu} \cdot \mathbf{e}_{P'}^{*} | \Psi_{+P+P'}(t') \rangle$$
$$= \sum_{p,q=\alpha,\beta} \Omega_{fq}^{P'} \Omega_{qg}^{P} \Omega_{gp}^{\overline{P}} \Omega_{pf}^{\overline{P'}} \mathrm{e}^{-\mathrm{i}\omega_{qp}T}, \tag{4.48}$$

which is identical to equation (4.44*a*). Here, just as in equation (4.44*c*), we have used the fact that $\mathrm{erf}(\mathrm{i}x)$ is purely imaginary for a real valued x.

Note that equations (4.44*a*)–(4.44*c*) are all real valued because each of the $p = q$ terms is real, and for every $p \neq q$ term, its complex conjugate is also included in the summation.

From this calculation we gain several insights:

- Even though S_{ESA} is always *positive* and S_{SE} is always *negative* for every value of T (regardless of the model), S_{GSB} happens to be a negative constant background in the vibrationless coupled dimer. The GSB contribution is independent of T because it features the population $|g\rangle\langle g|$ in the waiting time, which undergoes a trivial evolution in Liouville space. No coherence in the GSM is available because $|g\rangle$ is the only state in this particular model.
- S_{GSB} is a *negative* contribution to absorption in this model. We can understand this as follows. Consider the leftover GSM wavepacket after P acts but before P' arrives (equations (4.42) and (4.47)),

$$\mathbb{P}_{\text{GSM}}|\Psi\rangle = |g\rangle\langle g|(|\Psi_0\rangle + |\Psi_{PP}\rangle)$$
$$= |g\rangle \left(1 - \sum_{q=\alpha,\beta} \frac{|\Omega_{qg}^{P}|^2}{2} \right), \tag{4.49}$$

  which gives a GSM population that is less than before P arrived. In this example, $\mathbb{P}_{\text{GSM}}|\Psi_{PP}\rangle$ is a second-order wavefunction which comes with the opposite sign ($(\mathrm{i})(\mathrm{i}) = -1$) compared to the background wavefunction $|\Psi_0\rangle$. In the GSB literature, $|\Psi_{PP}\rangle$ is sometimes termed a 'hole' wavefunction [1, 10]. This represents a decrease in absorption due to having a smaller GSM population. This, in turn, manifests in a weaker absorption of P' compared to the linear absorption of P' had P not previously acted on the sample. Hence, for the ideal coupled dimer case, $S_{\text{GSB}} < 0$. We emphasize, however, that neither the T-independence nor the negativity are requirements for S_{GSB} in general, although for QPT purposes, we will try to find conditions that allow for both (see next chapter).
- S_{ESA} and S_{SE} both report on SEM dynamics as a function of the waiting time T, which can be sampled experimentally. Being the squared amplitude of a wavepacket, we can write equation (4.44*a*) as,

$$S_{\text{ESA}}(T) = \left(c_{\alpha}^{\text{ESA}}\right)^2 + \left(c_{\beta}^{\text{ESA}}\right)^2 + 2\left(c_{\alpha}^{\text{ESA}}\right)\left(c_{\beta}^{\text{ESA}}\right)\cos(\omega_{\alpha\beta}T), \tag{4.50}$$

where we have combined products of transition amplitudes into the real constants $c_{\alpha}^{\mathrm{ESA}}$ and c_{β}^{ESA}. An analogous equation for SE holds. The direct terms reflect the (trivial) dynamics of the populations $\mathrm{e}^{-\mathrm{i}\omega_{\alpha\alpha}T} = \mathrm{e}^{-\mathrm{i}\omega_{\beta\beta}T} = 1$, whereas the cross term reveals the coherence dynamics $2\Re \mathrm{e}^{-\mathrm{i}\omega_{\alpha\beta}T} = 2\cos(\omega_{\alpha\beta}T)$. A similar analysis can be made for $S_{\mathrm{SE}}(T)$. Since $S_{\mathrm{GSB}}(T)$ appears only as a T-independent background in this example, we may identify $S_{PP'}(T)$ as a direct reporter of SEM dynamics as a function of T. We investigate this issue further and generalize it to the context of dimers with coupling to vibrations in the next chapters.

Example 7. A witness to distinguish between electronic and vibrational coherences

We now show that the formalism above can be used to distinguish electronic and vibrational coherences in pump–probe spectroscopy [14].

Oscillations in a PP' signal $S_{PP'}(T)$ are associated with superpositions of vibronic states that are prepared by P. It is of particular interest to test whether these coherences are superpositions of vibrational states in a single electronic PES or superpositions of different electronic states [5, 12, 14]. These correspond to qualitatively different quantum phenomena and provide different physical mechanisms for coherent transport, which might be relevant to light harvesting in molecular aggregates. Vibrational superpositions within a single PES are expected to last longer than superpositions related to wavepackets in different PES, especially when they have very different geometries [6].

Consider the coupled dimer model (with vibrations), equations (2.1)–(2.7) but, just for this example, simplify it by setting, $V_a(\mathbf{R}) = \omega_a + V_{\mathrm{SEM}}(\mathbf{R})$, $V_b = \omega_b + V_{\mathrm{SEM}}(\mathbf{R})$, and $J(\mathbf{R}) = J$, that is, the two diabatic surfaces have the same geometry and their coupling does not depend on nuclear coordinates. This allows us to write H_{SEM} (equation (2.6)),

$$\begin{aligned} H_{\mathrm{SEM}} = {} & [T_N + V_{\mathrm{SEM}}(\mathbf{R})](|a\rangle\langle a| + |b\rangle\langle b|) \\ & + \omega_a|a\rangle\langle a| + \omega_b|b\rangle\langle b| + J(|a\rangle\langle b| + |b\rangle\langle a|). \end{aligned} \tag{4.51}$$

We can diagonalize the **R**-independent part by defining electronic states $|\alpha\rangle$ and $|\beta\rangle$, just as in section 3.2, and write,

$$H_{\mathrm{SEM}} = H_\alpha + H_\beta, \tag{4.52}$$

$$H_i = [T_N + \underbrace{\omega_i + V_{\mathrm{SEM}}(\mathbf{R})}_{\equiv V_i(\mathbf{R})}]|i\rangle\langle i| \quad \text{for} \quad i = \alpha, \beta. \tag{4.53}$$

Hence, $V_\alpha(\mathbf{R})$ and $V_\beta(\mathbf{R})$ become *adiabatic* PES: in this model, a wavepacket prepared in $|\alpha\rangle$ does not leave $|\alpha\rangle$, and the same with $|\beta\rangle$ [11].

For this model, we will show that in the limit of ultrashort broadband pulses P and P', oscillations in $S_{PP'}(T)$ are due to *electronic coherences*, that is, coherent superpositions between two electronic states. Outside the broadband limit, oscillations of

$S_{PP'}(T)$ arise due to both electronic and vibrational effects. For simplicity, we shall focus on the $S_{\mathrm{SE}}(T)$ piece of the signal, although the conclusions follow even after including the $S_{\mathrm{ESA}}(T)$ and S_{GSB} contributions into $S_{PP'}(T)$. It is important to remember that we are working under the Condon approximation, where $\boldsymbol{\mu}$ is independent of $\mathbf{R}$.

1. Derive an expression for $S_{\mathrm{SE}}(T)$; write it in terms of Franck–Condon (FC) overlaps. Use the vibronic basis $H_i|\nu_n^{(i)}\rangle = \omega_{i_n}|\nu_n^{(i)}\rangle$ for each of the relevant PES ($i = g, \alpha, \beta$). Assume that the initial state is $|\nu_n^{(g)}\rangle$, a vibrational eigenstate of H_g.
2. Take the limit of $S_{\mathrm{SE}}(T)$ when $\sigma_n \to 0$ using equation (3.21),

$$\lim_{\sigma_n \to 0} \tilde{\varepsilon}_n(\omega) = \eta. \tag{4.54}$$

Show that when only $|\alpha\rangle$ is bright, no oscillations are expected in the signal. Interpret this result. The limit above should be understood as the pulse being instantaneous compared to the dynamics in each excitation manifold, but larger than the inverse of the energy gaps between different manifolds, so that the RWA still holds.

Solution

1. In the chosen basis, the pump P takes amplitude from $|g\rangle|\nu_n^{(g)}\rangle$ in the GSM to $|j\rangle|\nu_r^{(j)}\rangle$ in the SEM with an amplitude $\Omega^P_{j_r, g_n} = \tilde{\varepsilon}_P(\omega_{j_r g_n})\boldsymbol{\mu}_{j_r g_n} \cdot \mathbf{e}_P$, where $j = \alpha, \beta$. After a time T, the probe P' takes $|j\rangle|\nu_r^{(j)}\rangle$ to $|g\rangle|\nu_s^{(g)}\rangle$ with an amplitude $\Omega^{\overline{P}'}_{g_s j_r} = \tilde{\varepsilon}^*_{P'}(\omega_{j_r g_s})\boldsymbol{\mu}_{g_s j_r} \cdot \mathbf{e}^*_{P'}$. In the Condon approximation, the transition dipoles can be expressed in terms of FC overlaps,

$$\Omega^P_{j_r g_n} = \tilde{\varepsilon}_P(\omega_{j_r g_n})\boldsymbol{\mu}_{jg} \cdot \mathbf{e}_P \langle \nu_r^{(j)}|\nu_n^{(g)}\rangle,$$

$$\Omega^{\overline{P}'}_{g_s j_r} = \tilde{\varepsilon}^*_{P'}(\omega_{j_r g_s})\boldsymbol{\mu}_{gj} \cdot \mathbf{e}^*_{P'} \langle \nu_s^{(g)}|\nu_r^{(j)}\rangle.$$

From this process, the asymptotic GSM wavepacket after the action of pulse P (equation (4.24)) is

$$\begin{aligned}\mathbb{P}_{\mathrm{GSM}}|\Psi_{PP'}\rangle = -\sum_{j=\alpha,\beta}\sum_{rs} &\left(\underbrace{|g\rangle|\nu_s^{(g)}\rangle}_{\text{Final state } |g\rangle|\nu_s^{(g)}\rangle}\right)\left(\underbrace{\boldsymbol{\mu}_{gj}\cdot\mathbf{e}^*_{P'}\tilde{\varepsilon}^*_{P'}(\omega_{j_r g_s})\overbrace{\langle\nu_s^{(g)}|\nu_r^{(j)}\rangle}^{\text{FC overlap}}}_{\text{Amplitude of 2nd transition}}\right)\\ &\times\left(\underbrace{(\boldsymbol{\mu}_{jg}\cdot\mathbf{e}_P)\tilde{\varepsilon}_P(\omega_{j_r g_n})\overbrace{\langle\nu_r^{(j)}|\nu_n^{(g)}\rangle}^{\text{FC overlap}}}_{\text{Amplitude of 1st transition}}\right)\\ &\times\left(\underbrace{\mathrm{e}^{\mathrm{i}\omega_{g_s}t_{P'}}}_{\text{Evolution in } |g\rangle|\nu_s^{(g)}\rangle}\right)\left(\underbrace{\mathrm{e}^{-\mathrm{i}\omega_{j_r}T}}_{\text{Evolution in } |j\rangle|\nu_r^{(j)}\rangle}\right)\left(\underbrace{\mathrm{e}^{-\mathrm{i}\omega_{g_n}t_P}}_{\text{Evolution in } |g\rangle|\nu_g^{(n)}\rangle}\right),\end{aligned} \tag{4.55}$$

which in turn gives, via equation (4.27*b*),

$$S_{\mathrm{SE}}(T) = -\sum_{j,j'=\alpha,\beta} (\boldsymbol{\mu}_{gj}\cdot\mathbf{e}^*_{P'})(\boldsymbol{\mu}_{jg}\cdot\mathbf{e}_P)(\boldsymbol{\mu}_{gj'}\cdot\mathbf{e}^*_P)(\boldsymbol{\mu}_{j'g}\cdot\mathbf{e}_{P'})$$
$$\times \sum_{rr's} \langle \nu_s^{(g)}|\nu_r^{(j)}\rangle\langle \nu_r^{(j)}|\nu_n^{(g)}\rangle\langle \nu_n^{(g)}|\nu_{r'}^{(j')}\rangle\langle \nu_{r'}^{(j')}|\nu_s^{(g)}\rangle$$
$$\times \tilde{\varepsilon}^*_{P'}(\omega_{j_r g_s})\tilde{\varepsilon}_P(\omega_{j_r g_n})\tilde{\varepsilon}^*_P(\omega_{j'_{r'} g_n})\tilde{\varepsilon}^*_{P'}(\omega_{j'_{r'} g_s})\mathrm{e}^{-\mathrm{i}\omega_{j_r j'_{r'}}T}. \quad (4.56)$$

In general, $S_{\mathrm{SE}}(T)$ is an oscillatory signal due to the phases $\mathrm{e}^{-\mathrm{i}\omega_{j_r j'_{r'}}T}$, which indicate coherent superpositions between the vibronic states $|j\rangle|\nu_r^{(j)}\rangle$ and $|j'\rangle|\nu_{r'}^{(j')}\rangle$ in the SEM, both within ($j = j'$) and between different ($j \neq j'$) surfaces.

2. In the broadband limit, but guaranteeing that RWA still holds, equation (4.55) can be simplified using equation (4.54). Performing the sums over r and s,

$$\mathbb{P}_{\mathrm{GSM}}|\Psi_{PP'}\rangle = -\eta^2 \sum_{j=\alpha,\beta} (\boldsymbol{\mu}_{gj}\cdot\mathbf{e}^*_{P'})(\boldsymbol{\mu}_{jg}\cdot\mathbf{e}_P)\underbrace{\Big(\sum_s \mathrm{e}^{\mathrm{i}\omega_{g_s}t_{P'}}|\nu_s^{(g)}\rangle\langle\nu_s^{(g)}|\Big)}_{=\mathrm{e}^{\mathrm{i}H_g t_{P'}}}$$
$$\times \underbrace{\Big(\sum_r \mathrm{e}^{-\mathrm{i}\omega_{j_r}T}|\nu_r^{(j)}\rangle\langle\nu_r^{(j)}|\Big)|\nu_n^{(g)}\rangle}_{=\mathrm{e}^{-\mathrm{i}H_j T}|\nu_n^{(g)}\rangle}\mathrm{e}^{-\mathrm{i}\omega_{g_n}t_P}$$
$$= -\eta^2 \sum_{j=\alpha,\beta} (\boldsymbol{\mu}_{gj}\cdot\mathbf{e}^*_{P'})(\boldsymbol{\mu}_{jg}\cdot\mathbf{e}_P)\mathrm{e}^{\mathrm{i}H_g t_{P'}}\underbrace{\mathrm{e}^{-\mathrm{i}H_j T}|\nu_n^{(g)}\rangle}_{\equiv|\psi_j(T)\rangle}\mathrm{e}^{-\mathrm{i}\omega_{g_n}t_P}, \quad (4.57)$$

where $|\psi_j(T)\rangle$ is the nuclear wavepacket in the electronic state j that results from the initial ground vibrational eigenstate impulsively launched onto the surface j. This wavepacket is a consequence of the Condon approximation, where under ideally broadband excitation, the vibrational state of the GSM gets copied with 100% fidelity into the SEM. Note that the wavepacket is not copied perfectly for narrowband excitation, where the different vibronic contributions of the launched wavepackets come weighted with the amplitude of the field at each transition energy.

Inserting equation (4.57) into equation (4.27*b*),

$$S_{\mathrm{SE}}(T) = -\sum_{j,j'=\alpha,\beta} \Omega^{\overline{P'}}_{gj}\Omega^{P}_{jg}\Omega^{\overline{P}}_{gj'}\Omega^{P'}_{j'g}\rho_{\mathrm{SEM},jj'}(T). \quad (4.58)$$

where we have used the definition of the transition amplitude (equations (3.20) and (3.22)) at the broadband limit (equation (4.54)).

Here, the *reduced electronic density matrix* of the SEM, $\rho_{\rm SEM}(T)$, in the $\{|\alpha\rangle, |\beta\rangle\}$ basis is given by,

$$\rho_{\rm SEM}(T) = \begin{pmatrix} \langle\psi_\alpha(T)|\psi_\alpha(T)\rangle & \langle\psi_\alpha(T)|\psi_\beta(T)\rangle \\ \langle\psi_\beta(T)|\psi_\alpha(T)\rangle & \langle\psi_\beta(T)|\psi_\beta(T)\rangle \end{pmatrix} \tag{4.59}$$

where the populations do not change because there is no coupling between the different adiabatic surfaces, and we have also ignored both radiative and non-radiative relaxation. We note that the *electronic coherences* are given by the overlap between two wavepackets, $\langle\psi_\alpha(T)|\psi_\beta(T)\propto e^{-i\omega_{\alpha\beta}T}$, with its magnitude remaining constant as a function of T. This is an artifact of assuming equal PES for $|a\rangle$ and $|b\rangle$, which is rare in practice. Typically, wavepackets in different PES will explore different configurations and their overlap will steadily decrease as T progresses, especially in the presence of a large number of nuclear degrees of freedom (such as in the case of a molecule embedded in a solvent), leading to electronic decoherence. Remarkably, in the broadband limit, even when the system is more complex than the one in example 6, the signal due to $S_{\rm SE}(T)$ can still be regarded as a linear combination of elements of a density matrix of the SEM [14].

We now draw attention to some limits. In the case when only $|\alpha\rangle$ is bright (i.e., $\boldsymbol{\mu}_{\beta g} = 0$), or when $|\alpha\rangle$ is the only electronic state in the SEM, as in an ideal monomer, $\rho_{\rm SEM}(T) = |\alpha\rangle\langle\alpha|$ and $S_{\rm SE}(T)$ does not depend on T, even though a large number of vibrational states may be excited. We can interpret this result as follows: in the broadband limit, the pulses contain every possible transition energy. In the Condon approximation, the transition dipole moment is independent of $\mathbf{R}$. These two observations imply that P' acts identically across every $\mathbf{R}$ to bring amplitude from $|a\rangle$ back to $|g\rangle$, and P' always 'catches' the *same* population $\langle\psi_\alpha(T)|\psi_\alpha(T)\rangle$ no matter where $|\psi_\alpha(T)\rangle$ is located in terms of nuclear coordinates. When both $|\alpha\rangle$ and $|\beta\rangle$ are bright, if there are any oscillations in $S_{\rm SE}(T)$, they must correspond to the electronic coherence $\langle\psi_\alpha(T)|\psi_\beta(T)\rangle$. This is the essence of a recently proposed witness for electronic coherence [14]. See figure 4.8 for a summary of the idea.

The analysis of $S_{\rm ESA}(T)$ can be done in an analogous way, except for the fact that in our coupled dimer model, $\boldsymbol{\mu}_{f\alpha} = \boldsymbol{\mu}_{\beta g}$, so if $|\alpha\rangle$ is the only bright state in the SEM, $S_{\rm ESA}(T) = 0$, which is by default a constant in T. $S_{\rm GSB}(T)$, on the other hand, is a probe for GSM dynamics. If there is a single PES in the GSM, as in our case, $S_{\rm GSB}(T)$ probes vibrational dynamics in $|g\rangle$ as a function of T. Vibrational coherences in $|g\rangle$ can be prepared and detected via $S_{\rm GSB}(T)$. However, in the broadband limit, $S_{\rm GSB}(T)$ becomes a constant background as a function of T, as we shall discuss in the next chapter. Hence, $S_{PP'}(T)$ altogether is a good witness for electronic coherence in the SEM, as we have described in [14]. In that article, we have also shown that more complicated measures based on two-dimensional spectroscopy do not readily distinguish vibrational from electronic coherences.

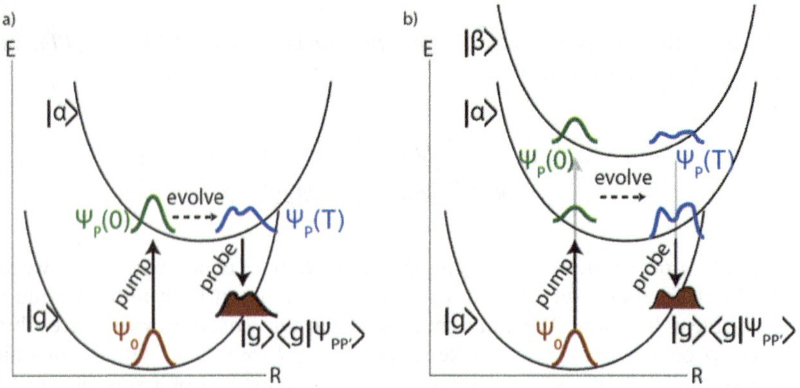

Figure 4.8. Illustration of broadband pump–probe spectra as a witness for coherent electronic oscillations. (*a*) For a monomer with a single excited state, in the broadband limit and assuming the Condon approximation, P and P' move amplitude in and out of the excited state regardless of the vibrational state; we observe this as leaving the shape of the vibrational wavepacket unperturbed by the pulses. Thus, the stimulated emission signal $S_{\mathrm{SE}}(T)$ is independent of the vibrational evolution during the waiting time T. (*b*) For a dimer with two excited states $|\alpha\rangle$, $|\beta\rangle$, even in the Condon approximation and broadband limit, the two single-excitation states produce two distinct interfering quantum pathways. The signal $S_{SE}(T)$ then oscillates as long as electronic coherence is maintained between $|\alpha\rangle$ and $|\beta\rangle$. (Reprinted with permission from Yuen-Zhou J, Krich J J and Aspuru-Guzik A 2012 *J. Chem. Phys.* **136** 234501. Copyright 2012, American Institute of Physics.)

Bibliography

[1] Banin U, Bartana A, Ruhman S and Kosloff R 1994 Impulsive excitation of coherent vibrational motion ground surface dynamics induced by intense short pulses *J. Chem. Phys.* **101** 8461–81

[2] Bergsma J P, Berens P H, Wilson K R, Fredkin D R and Heller E J 1984 Electronic spectra from molecular dynamics: a simple approach *J. Phys. Chem.* **88** 612–9

[3] Biggs J D and Cina J A 2009 Calculations of nonlinear wave-packet interferometry signals in the pump–probe limit as tests for vibrational control over electronic excitation transfer *J. Chem. Phys.* **131** 224302

[4] Cho M 2009 *Two Dimensional Optical Spectroscopy* (Boca Raton, FL: CRC Press)

[5] Egorova D 2008 Detection of electronic and vibrational coherences in molecular systems by 2d electronic photon echo spectroscopy *Chem. Phys.* **347** 166–76

[6] Franco I and Brumer P 2012 Electronic coherence dynamics in trans-polyacetylene oligomers *J. Chem. Phys.* **136** 144501

[7] Heller E J 1978 Quantum corrections to classical photodissociation models *J. Chem. Phys.* **68** 2066–75

[8] Heller E J 1981 The semiclassical way to molecular spectroscopy *Acc. Chem. Res.* **14** 368–75

[9] Henriksen N E and Engel V 2001 Femtosecond pump–probe spectroscopy: A theoretical analysis of transient signals and their relation to nuclear wave-packet motion *Int. Rev. Phys. Chem.* **20** 93–126

[10] Mukamel S 1995 *Principles of Nonlinear Optical Spectroscopy* (Oxford: Oxford University Press)
[11] Tannor D J 2007 *Introduction to Quantum Mechanics: A Time Dependent Approach* (Mill Valley, CA: University Science Books)
[12] Turner D B, Wilk K E, Curmi P M G and Scholes G D 2011 Comparison of electronic and vibrational coherence measured by two-dimensional electronic spectroscopy *J. Phys. Chem. Lett.* **2** 1904–11
[13] Yan Y J and Mukamel S 1990 Femtosecond pump–probe spectroscopy of polyatomic molecules in condensed phases *Phys. Rev.* A **41** 6485–504
[14] Yuen-Zhou J, Krich J J and Aspuru-Guzik A 2012 A witness for coherent electronic vs vibronic-only oscillations in ultrafast spectroscopy *J. Chem. Phys.* **136** 234501

Chapter 5

Putting it all together: quantum process tomography and pump–probe spectroscopies

We will now tie all the previous results together and show how a sequence of PP' experiments can be used to perform QPT. So far, we have shown that a PP' signal $S_{PP'}$ yields information on the dynamics of the SEM both for coupled dimers in the absence of coupling to a vibrational bath $\mathscr{B}$ (example 6) or in the context of strictly adiabatic electronic dynamics (example 7). In this chapter, we will lift these restrictions and extend the discussion to the general coupled dimer, which, regardless of the electronic basis of $\mathscr{S}$, can feature transfers between different electronic states due to coupling to vibrations in $\mathscr{B}$. We will show that for ultra-short broadband pulses P and P', and for a given waiting time T, the signal may be represented as a linear combination of elements of the process matrix $\chi(T)$ of the SEM, which is the main object to reconstruct in QPT. Yet, this observation means that a *single* PP' time trace does not, in general, provide enough information to elucidate the SEM dynamics at the quantum state level, as the desired information is contained in a linear combination. In fact, a series of PP' experiments with a set of different pulse parameters is required in order to invert the spectroscopic data to obtain $\chi(T)$. Carefully selecting the data to be collected will be the essence of our QPT procedure for ultrafast experiments.

5.1 Broadband PP' spectra in terms of the process matrix

We start by considering $S_{PP'}(T)$ in the limit of *ideal* broadband P and broadband P' cases ($\varepsilon_n = \delta(t - t_n)$ or $\tilde{\varepsilon}_n(\omega) = \eta$ for $n = P, P'$, equation (4.54)), and recall that we are working under the Condon approximation. The initial state is again assumed to be $|\Psi_0(0)\rangle = |g\rangle|\nu_n^{(g)}\rangle$. The asymptotic wavepackets from equations (4.31)–(4.33), for the coupled dimer with vibrations, take the form,

$$|\Psi_{P'}\rangle = \mathrm{i}\sum_{i,q=\alpha,\beta} \mathrm{e}^{\mathrm{i}H_0 t_{P'}}|i\rangle\langle i|\mathrm{e}^{-\mathrm{i}H_0 T}|q\rangle|\nu_n^{(g)}\rangle\Omega_{qg}^{P}\mathrm{e}^{-\mathrm{i}\omega_{gn} t_P}, \tag{5.1}$$

doi:10.1088/978-0-750-31062-8ch5

$$|\Psi_{PP'}\rangle = -\mathrm{e}^{\mathrm{i}H_0 t_{P'}}|g\rangle \sum_{i,q=\alpha,\beta} \Omega_{gi}^{\overline{P'}} \langle i|\mathrm{e}^{-\mathrm{i}H_0 T}|q\rangle|\nu_n^{(g)}\rangle\Omega_{qg}^{P}\mathrm{e}^{-\mathrm{i}\omega_{g_n}t_P}$$
$$-\mathrm{e}^{\mathrm{i}H_0 t_{P'}}|f\rangle \sum_{i,q=\alpha,\beta} \Omega_{fi}^{P'} \langle i|\mathrm{e}^{-\mathrm{i}H_0 T}|q\rangle|\nu_n^{(g)}\rangle\Omega_{qg}^{P}\mathrm{e}^{-\mathrm{i}\omega_{g_n}t_P}, \quad (5.2)$$

$$|\Phi_{PPP'}\rangle = -\mathrm{i}\sum_{i,q=\alpha,\beta} \mathrm{e}^{\mathrm{i}H_0 t_{P'}}|i\rangle\Omega_{ig}^{P'}\mathrm{e}^{-\mathrm{i}\omega_{g_n}T}|\nu_n^{(g)}\rangle\frac{\Omega_{gq}^{\overline{P}}\Omega_{qg}^{P}}{2}\mathrm{e}^{-\mathrm{i}\omega_{g_n}t_P}. \quad (5.3)$$

As opposed to the analogous equations for the vibrationless coupled dimer (equations (4.41)–(4.42)), in equations (5.1)–(5.2) we allow for the possibility of nonzero amplitude transfers from, say, α to β, $\langle\beta|\mathrm{e}^{-\mathrm{i}H_0 T}|\alpha\rangle \neq 0$, since in general, there is no strictly adiabatic electronic basis $\{\alpha, \beta\}$ in the presence of a vibrational bath $\mathscr{B}$ coupling to the electronic system $\mathscr{S}$. On the other hand, equation (5.3) is quite similar to equation (4.43) for $|\Phi_{PPP'}\rangle$ because P takes $|g\rangle$ to $|q\rangle \in \{|\alpha\rangle, |\beta\rangle\}$ and 'immediately' brings it back to $|g\rangle$, without letting $|q\rangle$ transfer to another electronic state. During the waiting time T, the electronic state is in $|g\rangle$, which does not transfer to other electronic states, even in the presence of vibrations because of the large gap between the GSM and the SEM or the DEM, is too large to be matched by the energy of the vibrations. Next, P' arrives and excites $|g\rangle$ into $|i\rangle \in \{|\alpha\rangle, |\beta\rangle\}$. Importantly, the distinction between equations (5.3) and (4.43) is that once $|i\rangle$ is prepared, $\mathrm{e}^{\mathrm{i}H_0 t_{P'}}|i\rangle \neq \mathrm{e}^{\mathrm{i}\omega_i t_{P'}}|i\rangle$ for a general coupled dimer.

Let us evaluate $S_{\mathrm{ESA}}(T)$. This time, we start by performing a classical average of the signal over the various initial states $|g\rangle|\nu_n^{(g)}\rangle$, which are drawn from a thermal ensemble with Boltzmann probability $p_n = \exp(-\omega_{g_n}/k_B\mathscr{T})/\mathrm{Tr}\,[\exp(-H_g/k_B\mathscr{T})]$, where k_B is the Boltzmann constant and $\mathscr{T}$ is the temperature of the ensemble. Equation (4.34) then reads,

$$\begin{aligned}
S_{\mathrm{ESA}}(T) &= \sum_n p_n \mathrm{Tr}_{\mathscr{B}}[\mathbb{P}_{\mathrm{DEM}}|\Psi_{PP'}\rangle\langle\Psi_{PP'}|\mathbb{P}_{\mathrm{DEM}}] \\
&= \sum_n p_n \mathrm{Tr}_{\mathscr{B}}\left[\left(\sum_{iq}\Omega_{fi}^{P'}\langle i|\mathrm{e}^{-\mathrm{i}H_0 T}|q\rangle|\nu_n^{(g)}\rangle\Omega_{qg}^{P}\right)\right. \\
&\quad \left.\times\left(\sum_{pj}\Omega_{gp}^{\overline{P}}\langle\nu_n^{(g)}|\langle p|\mathrm{e}^{\mathrm{i}H_0 T}|j\rangle\Omega_{jf}^{\overline{P'}}\right)\right] \\
&= \sum_{ijpq}\Omega_{fi}^{P'}\Omega_{qg}^{P}\Omega_{gp}^{\overline{P}}\Omega_{jf}^{\overline{P'}}\chi_{ijqp}(T), \qquad (5.4)
\end{aligned}$$

where we have used the definition of $\chi(T)$ from equation (1.5) and identified the initial density matrix of $\mathscr{B}$ as $\rho_{\mathscr{B}}(0) = p_n|\nu_n^{(g)}\rangle\langle\nu_n^{(g)}|$. Similarly, the SE and GSB contributions to $S_{PP'}(T)$ are given by,

$$S_{\text{SE}}(T) = -\sum_{ijpq} \Omega^{\overline{P}'}_{gi} \Omega^{P}_{qg} \Omega^{\overline{P}}_{gp} \Omega^{P'}_{jg} \chi_{ijqp}(T), \tag{5.5}$$

$$S_{\text{GSB}}(T) = -\sum_{ip} \Omega^{P'}_{ig} \Omega^{\overline{P}}_{gp} \Omega^{P}_{pg} \Omega^{\overline{P}'}_{gi}, \tag{5.6}$$

where $i, j, p, q \in \{\alpha, \beta\}$. Equations (5.4)–(5.6) constitute the core of this book and show the linear relation between a broadband PP' signal and the process matrix $\chi(T)$ for the electronic system in the SEM [22, 23]. The DS-FDs associated with each of these contributions are depicted in figure 5.1, which is a generalization of figure 4.7. In fact, these equations generalize equations (4.44a)–(4.44c) to situations where the bath $\mathscr{B}$ induces transfers between populations and coherences of the system $\mathscr{S}$. The signal PP' is a reporter on $\chi(T)$ since P and P' prepare and measure states in the SEM. Less trivial is the fact that the signals only report on the reduced state of the electronic system $\mathscr{S}$ and not of the entire molecule including vibrations. This can be traced back to a combination of the use of broad bandwidth as well as the Condon approximation.

Equation (5.1) describes the situation where the nuclear wavepacket $|\nu_n^{(g)}\rangle$ is promoted from $|g\rangle$ to $|q\rangle$ by P and after time T, via free evolution generated by H_{SEM}, ends up in electronic state $|i\rangle$, which may be different from $|q\rangle$. The information of this state gets transferred via P' into a superposition of wavepackets in $|f\rangle$ and $|g\rangle$ (equation (5.2)), which give $S_{\text{ESA}}(T)$ and $S_{\text{SE}}(T)$, respectively. These contributions depend on products of four dipole matrix elements associated with the different transitions between the excitation manifolds. Recall from chapter 1 that QPT is

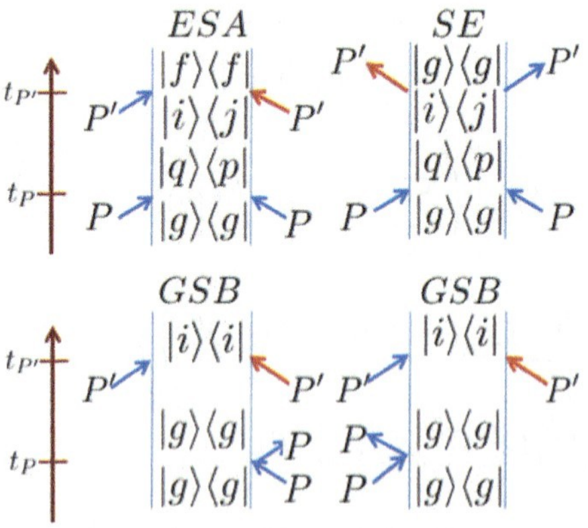

Figure 5.1. DS-FDs for a general broadband PP' signal. The ESA and SE pathways report on $\chi_{ijqp}(T)$: if $|q\rangle\langle p|$ is the initial state after the waiting time, what is the probability amplitude that the final state after the waiting time T will be $|i\rangle\langle j|$? These pathways differ from the ones in figure 4.7, where $|q\rangle\langle p|$ stays as $|q\rangle\langle p|$ because no bath is present that can induce a transfer to $|i\rangle\langle j|$. The GSB pathways provide a time-independent background as long as both pulses are sufficiently broadband.

well-defined if the initial state of the bath $\mathscr{B}$ is identical for all initial electronic system $\mathscr{S}$ states. In this experiment, at $T = 0^+$ (immediately after the action of P), the nuclei can be considered to be in state $\rho_B(0)$ for every initial state $|q\rangle\langle p|$ of the electrons. This is due to the Condon approximation and the extremely short (impulsive) perturbation. In this limit, the reduced electronic dynamics is given by equation (1.2).

We note that $S_{\text{GSB}}(T)$, being in general a probe for GSM dynamics, is again in this case independent of T, which can be understood, as in example 7, by considering one member of the initial thermal ensemble, $|\Psi_0(0)\rangle = |g\rangle|\nu_n^{(g)}\rangle$. In the broadband limit and Condon approximation, after P acts, $\mathbb{P}_{\text{GSM}}|\Psi_{PP}\rangle \propto |\nu_n^{(g)}\rangle|g\rangle = |\Psi_0\rangle$ (equation (4.37)); that is, P takes $|\nu_g^{(n)}\rangle$ to the SEM and instantaneously brings it back down to the GSM before the shape of the wavepacket can be distorted by evolution under H_{SEM}. Hence, $\mathbb{P}_{\text{GSM}}|\Psi_{PP}\rangle$ is proportional to $\Omega_{gq}^{\bar{P}}\Omega_{qg}^{P}/2$, because the same transition is probed in the two steps. Subsequently, since $|\Psi_0\rangle$ is an eigenstate of H_0, it remains stationary. $|\Psi_{P'}\rangle$ and $|\Phi_{PPP'}\rangle$ are the result of P' acting on $|\Psi_0(t)\rangle$ and $\mathbb{P}_{\text{GSM}}|\Psi_{PP}(t)\rangle \propto |\Psi_0(t)\rangle$, respectively, and therefore will be scaled copies of each other. Hence $\langle\Psi_{P'}|\Phi_{PPP'}\rangle$ are independent of T, and so is $S_{\text{GSB}}(T)$. This is true when spontaneous emission from the SEM is negligible, generally up to the nanosecond timescale. Importantly, once the states in the SEM decay back to the GSM, $S_{\text{GSB}}(T)$ vanishes since it corresponds to a differential signal (see example 5). This argument remains valid upon averaging over the initial thermal ensemble.

Although equations (5.4)–(5.6) have been derived in the ideal broadband limit, we will argue in the next section that they also hold in a quasi-broadband regime where the pulses are long enough that their spectrum is energetically selective to *electronic* transitions, but short enough that the vibrational bath $\mathscr{B}$ does not evolve significantly during the time windows in which the pulses act. This observation will provide a way to carry out QPT using ultrafast spectroscopy.

Example 8. Invariance of the broadband PP' signal under change of basis of the SEM

Show that each of the contributions of equations (5.4)–(5.6) is invariant under change of electronic basis in the SEM [22].

Solution

Let $|i\rangle = \sum_{i'} V_{ii'}|i'\rangle$ be the change of basis from the unprimed to the primed basis, where $\sum_j V_{ij}V_{i'j}^* = \delta_{ii'}$, that is, V is a unitary matrix. The process matrix transforms as,

$$\begin{aligned}\chi_{ijqp}(T) &= \mathrm{Tr}_B[\langle i|U(T)(|q\rangle\langle p| \otimes \rho_B(0))U^\dagger(T)|j\rangle] \\ &= \sum_{i''j''q''p''} V_{ii''}^* V_{qq''} V_{pp''}^* V_{jj''}\ \mathrm{Tr}_B[\langle i''|U(T)(|q''\rangle\langle p''| \otimes \rho_B(0))U^\dagger(T)|j''\rangle] \\ &= \sum_{i''j''q''p''} V_{ii''}^* V_{qq''} V_{pp''}^* V_{jj''}\chi_{i''j''q''p''}(T). \qquad (5.7)\end{aligned}$$

In the broadband limit where $\tilde{\varepsilon}_n(\omega) = \eta$ (equation (4.54)), $\Omega_{ij}^n = \eta \boldsymbol{\mu}_{ij} \cdot \mathbf{e}_n$. This implies that $\Omega_{ij}^n = \sum_{i'} V_{ii'}^* \Omega_{i'j}$ if i is in the SEM or $\Omega_{ij}^n = \sum_{j'} \Omega_{ij'} V_{jj'}$ if j is in the SEM. Hence, $S_{\mathrm{ESA}}(T)$ transforms as,

$$\begin{aligned} S_{\mathrm{ESA}}(T) &= \sum_{ijpq} \Omega_{fi}^{P'} \Omega_{qg}^{P} \Omega_{gp}^{\overline{P}} \Omega_{jf}^{\overline{P}'} \chi_{ijqp}(T) \\ &= \sum_{ijpq} \sum_{i'j'q'p'} \sum_{i''j''p''q''} \Omega_{fi'}^{P'} \Omega_{q'g}^{P} \Omega_{gp'}^{\overline{P}} \Omega_{j'f}^{\overline{P}'} \\ &\quad \times V_{ii'} V_{ii''}^* V_{qq''} V_{qq'}^* V_{pp'} V_{pp''}^* V_{jj''} V_{jj'}^* \chi_{i''j''q''p''}(T) \\ &= \sum_{i'j'p'q'} \Omega_{fi'}^{P'} \Omega_{q'g}^{P} \Omega_{gp'}^{\overline{P}} \Omega_{j'f}^{\overline{P}'} \chi_{i'j'q'p'}(T). \end{aligned} \tag{5.8}$$

Similarly,

$$S_{\mathrm{SE}}(T) = -\sum_{i'j'p'q'} \Omega_{gi'}^{\overline{P}'} \Omega_{q'g}^{P} \Omega_{gp'}^{\overline{P}} \Omega_{j'g}^{P'} \chi_{i'j'q'p'}(T), \tag{5.9}$$

$$S_{\mathrm{GSB}}(T) = -\sum_{i'p'} \Omega_{i'g}^{P'} \Omega_{gp'}^{\overline{P}} \Omega_{p'g}^{P} \Omega_{gi'}^{\overline{P}'}. \tag{5.10}$$

The structure of equations (5.4)–(5.10) is preserved upon change of basis. Even though the signals remain the same, the dynamics will be attributed to different elements of $\chi(T)$, depending on the basis in use.

We now demonstrate the utility of PP' spectroscopy to detect electronic coherences taking realistic values for pulse durations. We consider pulses whose full-width half maximum (FWHM) in the intensity $|\varepsilon_n(t - t_n)|^2$ is FWHM $= 2\sqrt{\ln 2}\sigma_n \approx 7$–15 fs), which are sufficiently broadband for the prototypical organic dyes we are studying. This idea is a generalization of the witness for electronic versus vibrational coherences presented in example 7. Here, as opposed to the situation in the mentioned example, we consider situations where the shape of the PES associated with $|\alpha\rangle$ and $|\beta\rangle$ need not be identical, hence allowing for a variety of transfers between populations and coherences. The details of the computational simulations can be found in [22]. Figure 5.2 shows, from left to right, calculated absorption spectra for a monomer with a single excited electronic state $|\alpha\rangle$, a 'coherent' dimer with strong J coupling between chromophores and an 'incoherent' dimer with weak J. The vibrations for each molecule have been assumed to be harmonic as in example 2. Besides homogeneous broadening due to vibrations, the spectra have been inhomogeneously broadened with site energy disorder to simulate a realistic ensemble. We have also plotted the frequency domain pulse power spectra on top of the spectra, showing that they cover all the transitions almost uniformly (this is the physical criterion of a broadband pulse for our purposes).

Figure 5.3, top panel, shows inhomogeneously broadened $S_{PP'}(T)$ corresponding to these simulations. The bottom panel shows a few time traces of elements of $\chi(T)$

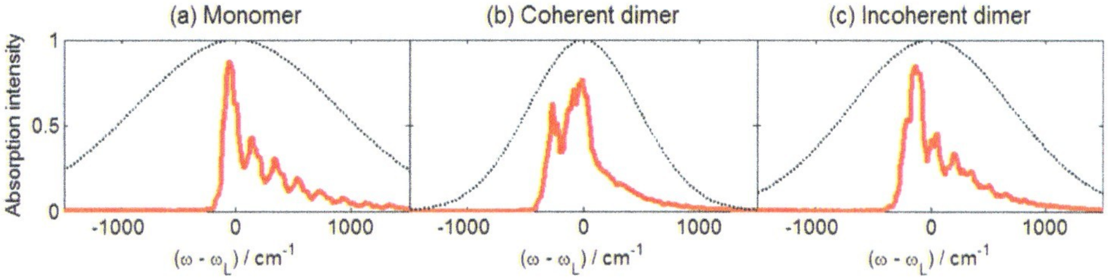

Figure 5.2. Inhomogeneously broadened absorption spectra of chromophores (solid red) with pulse spectral profiles $|\tilde{\varepsilon}_P(\omega)|^2 = |\tilde{\varepsilon}_n(\omega)|^2$ on top (dotted black). Since there is only one electronic excited state in the monomer, the peaks in its absorption spectrum must correspond to vibrations. This conclusion does not necessarily follow for the dimers because of the presence of multiple excited states in the SEM. (Reprinted with permission from Yuen-Zhou J, Krich J J and Aspuru-Guzik A 2012 *J. Chem. Phys.* **136** 234501. Copyright 2012, American Institute of Physics.)

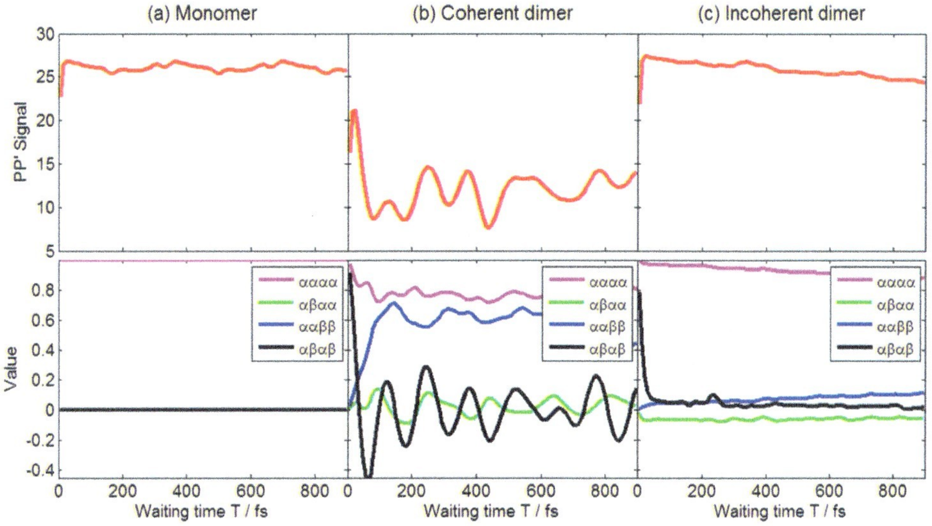

Figure 5.3. (Top) Broadband PP' spectra as a function of waiting time T as a witness for electronic coherences. The small oscillations in (a) and (c) are due to finite pulse durations. (Bottom) The witness is a linear combination of elements of the process matrix $\chi(T)$. Traces of a few representative elements of $\chi(T)$ are shown for illustration. (Reprinted with permission from Yuen-Zhou J, Krich J J and Aspuru-Guzik A 2012 *J. Chem. Phys.* **136** 234501. Copyright 2012, American Institute of Physics.)

obtained from computational simulations using wavepackets (see chapter 6). In order to plot the different elements of $\chi(T)$, we have chosen the electronic eigenbasis of H_S, given by equation (2.17), but, as shown in example 8, the choice of basis is arbitrary. However, we note that the dynamics of the elements of $\chi(T)$ translates into the dynamics of $S_{PP'}(T)$ in the broadband limit, as equations (5.4)–(5.6) show. The oscillations of the PP' signal of the coherent dimer are much larger in magnitude than those of the monomer and the incoherent dimer. In the monomer, as we have explained in example 7 and figure 4.8, oscillations vanish in the ideal broadband limit. Since the pulses we have used in the simulation are broadband, but still of finite bandwidth, as can be seen in figure 5.2, remnants of oscillations survive. Ways to calibrate the magnitude of these oscillations have been proposed in [22]. Either way, in the Condon approximation, oscillations in frequency-integrated broadband PP' spectra directly reflect *electronic* dynamics (instead of vibronic dynamics as a whole). Hence, oscillations in these measurements are strong signatures for dynamical electronic coherences, in some basis.

5.2 Performing QPT using PP' data

We now discuss how to extract $\chi(T)$ for the SEM. For PP' spectroscopy, equations (5.4)–(5.6) show that the signal contains information about $\chi(T)$. In order to obtain $\chi(T)$, and hence achieve QPT, it is necessary to generate an invertible system of linear equations from $S_{PP'}(T)$ data. This requires at least the same number of linearly independent measurements as unknowns. From example 1, part 4-d, we know that for a d-dimensional Hilbert space, $\chi(T)$ contains $d^4 - d^2$ real independent parameters. This means that for a SEM of two electronic states, 12 linearly independent measurements are needed.

From equations (5.4)–(5.6), in broadband PP' spectroscopy, the only possibility to generate the system of equations is by manipulating the polarization of the fields with respect to the dipoles. However, under isotropic averaging of the signals (molecules have different orientations in solution, and on the timescales of the experiment can be considered to be rotationally static), this possibility yields at most three linearly independent measurements (for an introduction to isotropic averaging, see appendix F). Hence, the reconstruction of $\chi(T)$ for an arbitrary coupled dimer is an impossible task with frequency-integrated PP' data from broadband pulses alone. However, partial QPT is possible with only polarization control and, in fact, can be realized using a variant of PP' spectroscopy, two-dimensional electronic spectroscopy [21] (see also appendices E and F). In this book, we take a different approach, focusing on carrying out total QPT with PP' spectroscopy by exploiting additional parameters of the pulses, such as their frequency characteristics.

Within the dipole approximation (equation (3.2)), the available pulse parameters are polarizations, phases, frequencies and time delays. These properties might be useful to achieve selectivity in the creation and detection of states in Liouville space. So far, phases have disappeared from our expressions of $S_{PP'}(T)$ and time delays have not been exploited. On the other hand, although the derivation of equations (5.4)–(5.6) depends on δ pulses, an appropriate argument on separation of timescales allows us to exploit frequency control to achieve QPT. This is our goal in this chapter.

In the rest of the book, we assume that the dimer SEM is composed of two electronic states, $|\alpha\rangle$ and $|\beta\rangle$, which are sufficiently separated in energy that the linear absorption spectrum consists of two approximately nonoverlapping vibronic progressions. Then, one can devise pulses such that their frequency spectrum $\tilde{\varepsilon}_n(\omega)$ is broadband compared to the vibrational energies, but still centered at one electronic

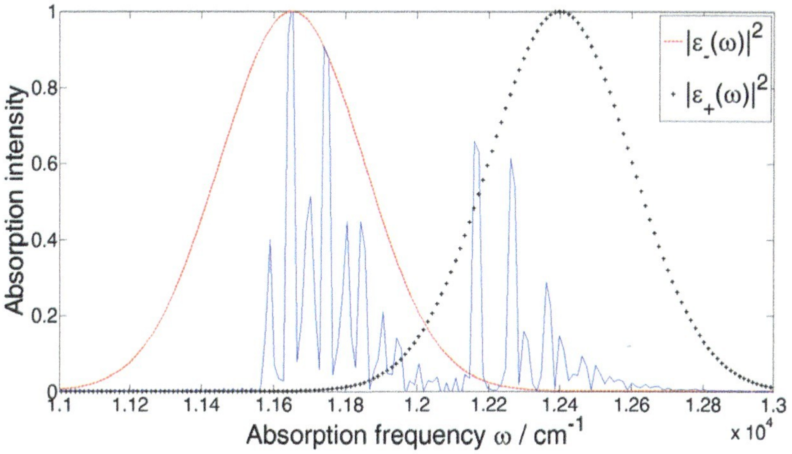

Figure 5.4. Absorption spectrum of a model dimer with well-separated singly-excited electronic states $|\alpha\rangle$ and $|\beta\rangle$ (solid line). Two optical pulses with amplitudes shown with $-$ and $+$ allow selective excitation of each electronic state (together with its vibronic progression). This two-pulse toolbox allows for selective initial-state preparation and final-state detection for QPT.

state or another. Then we can write expressions like $\tilde{\varepsilon}_n(\omega_{\alpha g})$, $\tilde{\varepsilon}_n(\omega_{\beta g})$, $\tilde{\varepsilon}_n(\omega_{f\alpha})$, $\tilde{\varepsilon}_n(\omega_{f\beta})$, which associate a single pulse amplitude with a given *electronic* transition (as opposed to amplitudes for each vibronic transition). In the time domain, this situation refers to a quasi-impulsive limit where pulses are selective in frequency, but still short compared to the dynamics induced by the vibrational bath $\mathscr{B}$. In this situation, equations (5.4)–(5.6) remain valid, but we note that the transition amplitudes between *electronic* states $\Omega_{ij}^n = \tilde{\varepsilon}_n(\omega_{ij})\boldsymbol{\mu}_{ij}\cdot\mathbf{e}_n$ have magnitudes controlled by $\tilde{\varepsilon}_n(\omega_{ij})$ [15, 21–23]. For simplicity, we also assume that

$$\omega_{\alpha g} \approx \omega_{f\beta}, \tag{5.11a}$$

$$\omega_{\beta g} \approx \omega_{f\alpha}, \tag{5.11b}$$

and that the transition dipole moments are all known, although, as explained later (see footnote 2 in example 9), this requirement may be lifted. In any case, these dipoles can be determined self-consistently with QPT [21] or via techniques such as x-ray crystallography combined with *ab initio* modeling [13].

Consider two types of pulses of different color (central frequency), denoted by $+$ and $-$, such that (see figure 5.4) [23],

$$\tilde{\varepsilon}_-(\omega_{\alpha g}) = \tilde{\varepsilon}_+(\omega_{\beta g}) \gg \tilde{\varepsilon}_-(\omega_{\beta g}) = \tilde{\varepsilon}_+(\omega_{\alpha g}) \tag{5.12}$$

We can choose to selectively prepare $|\alpha\rangle\langle\alpha|$ $(P = -)$ or $|\beta\rangle\langle\beta|$ $(P = +)$. Then we can selectively detect the final state $|\alpha\rangle\langle\alpha|$ via SE and $|\beta\rangle\langle\beta|$ via ESA $(P' = -)$, or $|\beta\rangle\langle\beta|$ via SE and $|\alpha\rangle\langle\alpha|$ via ESA $(P' = +)$. Hence, frequency-selective PP' data yield

Table 5.1. List of PP' experiments that obtain different elements of $\chi(T)$.

P	P'	Elements of $\chi(T)$
$-$	$-$	$\chi_{\alpha\alpha\alpha\alpha}(T)$, $\chi_{\beta\beta\alpha\alpha}(T)$
$-$	$+$	$\chi_{\alpha\alpha\alpha\alpha}(T)$, $\chi_{\beta\beta\alpha\alpha}(T)$
$+$	$-$	$\chi_{\alpha\alpha\beta\beta}(T)$, $\chi_{\beta\beta\beta\beta}(T)$
$+$	$+$	$\chi_{\alpha\alpha\beta\beta}(T)$, $\chi_{\beta\beta\beta\beta}(T)$

linear combinations of $\chi(T)$ that trace dynamics of populations. By carrying out four different $S_{PP'}(T)$ experiments by letting P, $P' \in \{+, -\}$, we may obtain signals that depend on only two elements of $\chi(T)$ at a time (table 5.1). For instance, by letting $P = -$ and $P' = +$, we see that,

$$S_{\mathrm{ESA}}(T) = \Omega^{P'}_{f\alpha}\Omega^{P}_{\alpha g}\Omega^{\overline{P}}_{g\alpha}\Omega^{\overline{P}'}_{\alpha f}\chi_{\alpha\alpha\alpha\alpha}(T), \tag{5.13}$$

$$S_{\mathrm{SE}}(T) = -\Omega^{\overline{P}'}_{g\beta}\Omega^{P}_{\alpha g}\Omega^{\overline{P}}_{g\alpha}\Omega^{P'}_{\beta g}\chi_{\beta\beta\alpha\alpha}(T), \tag{5.14}$$

$$S_{\mathrm{GSB}}(T) = -\Omega^{P'}_{\beta g}\Omega^{\overline{P}}_{g\alpha}\Omega^{P}_{\alpha g}\Omega^{\overline{P}'}_{g\beta}. \tag{5.15}$$

Via standard linear algebra, one can extract the elements of $\chi(T)$ in table 5.1 from the resulting signals by inverting a matrix consisting of products of dipoles [23]. This is an intuitive result and can be thought of in terms of chemical kinetics. P creates a specific population $|\alpha\rangle\langle\alpha|$ or $|\beta\rangle\langle\beta|$ and P' selectively detects them. Furthermore, this set is actually redundant. If we consider the constraint of trace preservation (equation (1.18)), we may write,

$$\chi_{\beta\beta\alpha\alpha}(T) = 1 - \chi_{\alpha\alpha\alpha\alpha}(T), \tag{5.16}$$

$$\chi_{\alpha\alpha\beta\beta}(T) = 1 - \chi_{\beta\beta\beta\beta}(T), \tag{5.17}$$

so that we only have two unknowns and we only need two experiments, say, the first and the fourth one in table 5.1. However, as we discuss in the following examples, we need other experiments to detect elements of $\chi(T)$ involving coherences.

Example 9. Adapting a PP' experiment into full QPT

In this example, we show that a few changes to the standard PP' experiment yield full QPT. The key idea is to consider a four-pulse experiment (four-wave mixing) where two collinear pulses together act as the pump P and two collinear pulses together act as the probe P', and importantly, with definite phase relationships between the two pumps and between the two probes, respectively. The purpose of this setup is to exploit the phases to turn certain contributions to the regular PP' signal on and off, and hence isolate elements of $\chi(T)$ selectively.

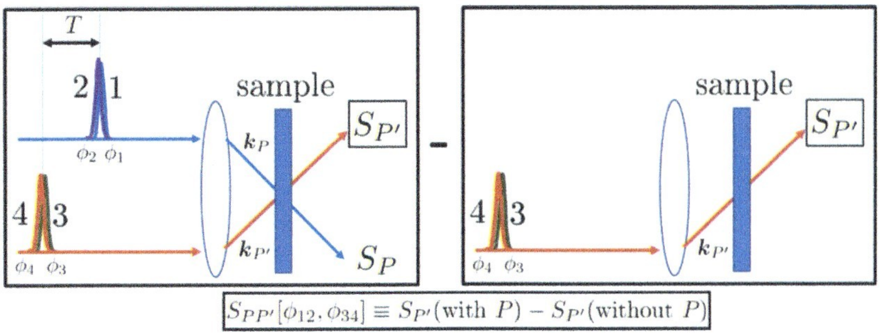

Figure 5.5. *PP′* spectroscopy with two pulse pairs. Pulses 1 and 2 act as the pump *P* and pulses 3 and 4 as the probe *P′*. The resulting *PP′* signal depends on the phases ϕ_{12} and ϕ_{34}, and on the spectral components of the four pulses, which can be independently manipulated. This setup provides enough control knobs, which can be manipulated accordingly to suppress or enhance Liouville pathways in the dynamics of the SEM, and hence, carry out full QPT of the SEM.

More precisely, let the four-pulse electric field be described by,

$$\boldsymbol{\varepsilon}(\mathbf{r}, t) = \sum_{n=1}^{4} \left[\varepsilon_n(t - t_n) \mathrm{e}^{\mathrm{i}\mathbf{k}_n \cdot \mathbf{r} + \mathrm{i}\phi_n} \mathbf{e}_n + \text{c.c.} \right]. \tag{5.18}$$

Here, the pulses 1 and 2 are collinear ($\hat{\mathbf{k}}_1 = \hat{\mathbf{k}}_2$), and pulses 3 and 4 are also collinear but along a different direction ($\hat{\mathbf{k}}_3 = \hat{\mathbf{k}}_4 \neq \hat{\mathbf{k}}_1$). Pulses 1 and 2 overlap in time and act as the pump ($t_1 = t_2 \equiv t_P$), and similarly for 3 and 4 as the probe ($t_3 = t_4 \equiv t_{P'}$). The polarization of the pulses is constant throughout, $\mathbf{e}_n = \mathbf{e}$ for all n. The frequencies of all four pulses can be different, but we consider the case where they are close enough that we can approximate $\mathbf{k}_1 \approx \mathbf{k}_2 \equiv \mathbf{k}_P$ and $\mathbf{k}_3 \approx \mathbf{k}_4 \equiv \mathbf{k}_{P'}$, but $\mathbf{k}_P \neq \mathbf{k}_{P'}$ as the pulse pair directions are different.

Devise a strategy to extract all the elements of $\chi(T)$, and hence achieve QPT, using the signals $S_{PP'}(T)$ resulting from this setup. Use equations (5.11*a*) and (5.11*b*) and the toolbox of equations (5.12*a*) and (5.12*b*).

Hint: consider a series of *PP′* signals resulting from different pulse frequencies and phases ϕ_i. Figure 5.6 summarizes the idea of frequency control from each of the four pulses in order to isolate a signal that reports solely on a single element of $\chi(T)$, in particular, $\chi_{\beta\alpha\alpha\beta}(T)$.

Solution

We first rewrite equation (5.18) in the form of equation (3.8), namely, as the sum of an effective pump *P* and an effective probe *P′* pulses,

$$\boldsymbol{\varepsilon}(\mathbf{r}, t) = \left\{ \underbrace{\left[\varepsilon_1(t - t_P)\mathrm{e}^{\mathrm{i}\phi_{12}} + \varepsilon_2(t - t_P)\right]}_{\equiv \varepsilon_P(t - t_P)} \mathrm{e}^{\mathrm{i}\mathbf{k}_P \cdot \mathbf{r} + \mathrm{i}\overbrace{\phi_2}^{\equiv \phi_P}} \mathbf{e} + \text{c.c.} \right\}$$

$$+ \left\{ \underbrace{\left[\varepsilon_3(t - t_{P'})\mathrm{e}^{\mathrm{i}\phi_{34}} + \varepsilon_4(t - t_{P'})\right]}_{\equiv \varepsilon_{P'}(t - t_{P'})} \mathrm{e}^{\mathrm{i}\mathbf{k}_{P'} \cdot \mathbf{r} + \mathrm{i}\overbrace{\phi_4}^{\equiv \phi_{P'}}} \mathbf{e} + \text{c.c.} \right\}, \tag{5.19}$$

where we have identified the effective values of ε_P and $\varepsilon_{P'}$, and taken $\phi_P = \phi_2$ and $\phi_{P'} = \phi_4$. Importantly, as opposed to the standard PP' setting or the linear absorption configuration, the relative phases $\phi_{12} \equiv \phi_1 - \phi_2$ and $\phi_{34} \equiv \phi_3 - \phi_4$ remain in the expressions for ε_P and $\varepsilon_{P'}$, respectively, so they will affect the signals[1]. Substitution of ε_P and $\varepsilon_{P'}$ into equations (4.20)–(4.29*b*) yields

$$S_{PP}[\phi_{12}, \phi_{34}](T) = S_{\mathrm{ESA}}[\phi_{12}, \phi_{34}](T) + S_{\mathrm{SE}}[\phi_{12}, \phi_{34}](T) + S_{\mathrm{GSB}}[\phi_{12}, \phi_{34}](T), \tag{5.20}$$

where we have explicitly expressed the PP' signal as a function of the relative phases. Each term of equation (5.20) is given by

$$\begin{aligned} S_{\mathrm{ESA}}[\phi_{12}, \phi_{34}](T) &= \langle \Psi_{PP'} | \mathbb{P}_{\mathrm{DEM}} | \Psi_{PP'} \rangle \\ &= \langle \Psi_{13} | \mathbb{P}_{\mathrm{DEM}} | \Psi_{13} \rangle + \langle \Psi_{14} | \mathbb{P}_{\mathrm{DEM}} | \Psi_{14} \rangle + \langle \Psi_{23} | \mathbb{P}_{\mathrm{DEM}} | \Psi_{23} \rangle \\ &\quad + \langle \Psi_{24} | \mathbb{P}_{\mathrm{DEM}} | \Psi_{24} \rangle + 2\Re\Big\{ \mathrm{e}^{\mathrm{i}\phi_{12}} (\langle \Psi_{23} | \mathbb{P}_{\mathrm{DEM}} | \Psi_{13} \rangle \\ &\quad + \langle \Psi_{24} | \mathbb{P}_{\mathrm{DEM}} | \Psi_{14} \rangle) + \mathrm{e}^{\mathrm{i}\phi_{34}} (\langle \Psi_{14} | \mathbb{P}_{\mathrm{DEM}} | \Psi_{13} \rangle \\ &\quad + \langle \Psi_{24} | \mathbb{P}_{\mathrm{DEM}} | \Psi_{23} \rangle) + \mathrm{e}^{\mathrm{i}(\phi_{12}+\phi_{34})} \langle \Psi_{24} | \mathbb{P}_{\mathrm{DEM}} | \Psi_{13} \rangle \\ &\quad + \mathrm{e}^{\mathrm{i}(-\phi_{12}+\phi_{34})} \langle \Psi_{14} | \mathbb{P}_{\mathrm{DEM}} | \Psi_{23} \rangle \Big\}, \end{aligned} \tag{5.21}$$

$$\begin{aligned} S_{\mathrm{SE}}[\phi_{12}, \phi_{34}](T) &= -\langle \Psi_{PP'} | \mathbb{P}_{\mathrm{GSM}} | \Psi_{PP'} \rangle \\ &= -\Big(\langle \Psi_{13} | \mathbb{P}_{\mathrm{GSM}} | \Psi_{13} \rangle + \langle \Psi_{14} | \mathbb{P}_{\mathrm{GSM}} | \Psi_{14} \rangle + \langle \Psi_{23} | \mathbb{P}_{\mathrm{GSM}} | \Psi_{23} \rangle \\ &\quad + \langle \Psi_{24} | \mathbb{P}_{\mathrm{GSM}} | \Psi_{24} \rangle + 2\Re\Big\{ \mathrm{e}^{\mathrm{i}\phi_{12}} (\langle \Psi_{23} | \mathbb{P}_{\mathrm{GSM}} | \Psi_{13} \rangle \\ &\quad + \langle \Psi_{24} | \mathbb{P}_{\mathrm{GSM}} | \Psi_{14} \rangle) + \mathrm{e}^{\mathrm{i}\phi_{34}} (\langle \Psi_{13} | \mathbb{P}_{\mathrm{GSM}} | \Psi_{14} \rangle \\ &\quad + \langle \Psi_{23} | \mathbb{P}_{\mathrm{GSM}} | \Psi_{24} \rangle) + \mathrm{e}^{\mathrm{i}(\phi_{12}+\phi_{34})} \langle \Psi_{23} | \mathbb{P}_{\mathrm{GSM}} | \Psi_{14} \rangle \\ &\quad + \mathrm{e}^{\mathrm{i}(-\phi_{12}+\phi_{34})} \langle \Psi_{13} | \mathbb{P}_{\mathrm{GSM}} | \Psi_{24} \rangle \Big\} \Big), \end{aligned} \tag{5.22}$$

$$\begin{aligned} S_{\mathrm{GSB}}[\phi_{12}, \phi_{34}](T) &= -2\Re\{ \langle \Psi_{P'} | \Phi_{PPP'} \rangle \} \\ &= 2\Re\{ \langle \Psi_3 | \Phi_{113} \rangle + \langle \Psi_3 | \Phi_{223} \rangle + \langle \Psi_4 | \Phi_{114} \rangle + \langle \Psi_4 | \Phi_{224} \rangle \\ &\quad + \mathrm{e}^{\mathrm{i}\phi_{12}} (\langle \Psi_3 | \Phi_{123} \rangle + \langle \Psi_4 | \Phi_{124} \rangle) + \mathrm{e}^{-\mathrm{i}\phi_{12}} (\langle \Psi_3 | \Phi_{213} \rangle + \langle \Psi_4 | \Phi_{214} \rangle) \\ &\quad + \mathrm{e}^{\mathrm{i}\phi_{34}} (\langle \Psi_4 | \Phi_{113} \rangle + \langle \Psi_4 | \Phi_{223} \rangle) + \mathrm{e}^{-\mathrm{i}\phi_{34}} (\langle \Psi_3 | \Phi_{114} \rangle + \langle \Psi_3 | \Phi_{224} \rangle) \\ &\quad + \mathrm{e}^{\mathrm{i}(\phi_{12}+\phi_{34})} \langle \Psi_4 | \Phi_{123} \rangle + \mathrm{e}^{\mathrm{i}(-\phi_{12}+\phi_{34})} \langle \Psi_4 | \Phi_{213} \rangle \\ &\quad + \mathrm{e}^{\mathrm{i}(\phi_{12}+\phi_{34})} \langle \Psi_3 | \Phi_{124} \rangle + \mathrm{e}^{\mathrm{i}(-\phi_{12}+\phi_{34})} \langle \Psi_3 | \Phi_{214} \rangle \}. \end{aligned} \tag{5.23}$$

[1] On the other hand, relative phases between pump and probe pulses do not appear here, as expected from our previous calculations on *PP'* spectroscopy. For an experimentalist, this means that control of relative phases among the four pulses is not necessary. It is only important between pulses 1 and 2, and 3 and 4, pairwise.

Here, the asymptotic wavepackets $|\Psi_i\rangle$, $|\Psi_{ij}\rangle$ and $|\Phi_{ijk}\rangle$ for $i, j, k \in \{1, 2, 3, 4\}$ are defined in analogy to equations (4.20)–(4.25), being labeled with the pulses that create them. An easy way to understand the different phase factors is via the RWA (equation (3.25)): $\varepsilon_i(t - t_i)$ comes with $e^{i\mathbf{k}_i\cdot\mathbf{r}+i\phi_i}$ (equations (3.3) and (5.19)) so it excites a ket to a higher energy state, whereas $\varepsilon_i^*(t - t_i)$ comes with $e^{-i\mathbf{k}_i\cdot\mathbf{r}-i\phi_i}$ and de-excites a ket to a lower energy state. The opposite holds for perturbations on the bra. As an example, the term $\langle\Psi_{14}|\mathbb{P}_{\mathrm{DEM}}|\Psi_{23}\rangle$ in $S_{\mathrm{ESA}}(T)$ comes with the phase $e^{i(-\phi_1+\phi_2+\phi_3-\phi_4)}$ because it involves excitations in the ket with pulses 2 and 3, and excitations in the bra with pulses 1 and 4. Similarly, the term $\langle\Psi_{23}|\mathbb{P}_{\mathrm{GSM}}|\Psi_{14}\rangle$ comes with the phase $e^{i(\phi_1-\phi_2+\phi_3-\phi_4)}$ because it involves excitation and de-excitation of the ket with pulses 1 and 4, and excitation and de-excitation of the bra with pulses 2 and 3.

Equations (5.21)–(5.23) can be compactly organized as a Fourier sum,

$$S_{PP'}[\phi_{12}, \phi_{34}](T) = \sum_{n_{12}, n_{34}=0, \pm 1} \hat{S}[n_{12}, n_{34}](T)e^{i(n_{12}\phi_{12}+n_{34}\phi_{34})}, \tag{5.24}$$

where we have indexed the expansion coefficients $\hat{S}$ via the integer pairs $[n_{12}, n_{34}]$. Since $S_{PP'}$ is a real-valued quantity, the (complex) coefficients satisfy the symmetry $\hat{S}[n_{12}, n_{34}](T) = \{\hat{S}[-n_{12}, -n_{34}](T)\}^*$. There are only nine $[n_{12}, n_{34}]$ pairs to consider, namely, [0, 0], [1, 0], [−1, 0], [0, 1], [0, −1], [1, 1], [−1, −1], [1, −1] and [−1, 1], so, in principle, one may find the coefficients $\hat{S}[n_{12}, n_{34}]$ assuming that sufficiently many linearly independent equations of the form of equation (5.24), which can be generated from signals associated with different phases $[\phi_{12}, \phi_{34}]$. This procedure is known as *phase cycling* (PC) and it is a ubiquitous tool in nonlinear spectroscopy [6, 14, 16, 18, 20].

Let us pause to discuss the physical meaning of each of the components $\hat{S}[n_{12}, n_{34}](T)$. $\hat{S}[0, 0](T)$ is a signal that stems from wavepacket overlaps that would have occurred in a standard *PP′* experiment with only one pump pulse and one probe pulse, and, consistent with our previous discussions, does not depend on relative phases between any of the pulses. $\hat{S}[1, 0](T)$, $\hat{S}[-1, 0](T)$, $\hat{S}[0, 1](T)$ and $\hat{S}[0, -1](T)$ are signals corresponding to interferences among two pump pulses and one probe pulse, or among one pump pulse and two probe pulses, respectively, and depend on the relative phases between only two of the pulses at a time. Finally, $\hat{S}[1, 1](T)$, $\hat{S}[-1, -1](T)$, $\hat{S}[1, -1](T)$ and $\hat{S}[-1, 1](T)$ depend on interferences created by the two pump pulses *and* the two probe pulses, and hence, depend on the relative phases between two pulse pairs. Therefore, it is useful to think about the contributions due to $\hat{S}[0, 0](T)$ as 'direct terms' and the ones due to the rest of the components as 'cross terms.' One can imagine that the cross terms, which depend on all four pulses, will be particularly useful for QPT, since each pulse can be tuned independently, providing more selectivity than the simple two-color *PP′* experiment from the last chapter. In this example, we will extract $\hat{S}[-1, 1](T)$ via PC and use it to carry out QPT, although similar conclusions can be obtained by focusing on $\hat{S}[1, 1](T)$, $\hat{S}[-1, -1](T)$ or $\hat{S}[1, -1](T)$.

Table 5.2 shows an example of a 9 × 9 matrix of coefficients $\{e^{i(n_{12}\phi_{12}+n_{34}\phi_{34})}\}$ which serves our purposes for PC. In this matrix, the rows correspond to nine specific realizations of the signal $S_{PP'}[\phi_{12}, \phi_{34}]$ by varying the phases $[\phi_{12}, \phi_{34}]$, whereas the columns are associated with the integers $[n_{12}, n_{34}]$ that label the coefficients of the expansion in equation (5.24).

One can numerically check that the matrix is invertible, which means that this set of nine signals provides a system of nine linearly independent equations from which the

Table 5.2. Matrix of phases $e^{i(n_{12}\phi_{12}+n_{34}\phi_{34})}$ for phase cycling (PC).

$[\phi_{12}, \phi_{34}]\backslash[n_{12}, n_{34}]$	[0, 0]	[1, 0]	[−1, 0]	[0, 1]	[0, −1]	[1, 1]	[−1, −1]	[1, −1]	[−1, 1]
$[0, 0]$	1	1	1	1	1	1	1	1	1
$[0, \pi]$	1	1	1	−1	−1	−1	−1	−1	−1
$\left[0, \frac{\pi}{2}\right]$	1	1	1	i	−i	i	−i	−i	i
$\left[-\frac{\pi}{2}, 0\right]$	1	−i	i	1	1	−i	i	−i	i
$\left[-\frac{\pi}{2}, \pi\right]$	1	−i	i	−1	−1	i	−i	i	−i
$\left[-\frac{\pi}{2}, \frac{\pi}{2}\right]$	1	−i	i	i	−i	1	1	−1	−1
$\left[\frac{\pi}{2}, 0\right]$	1	i	−i	1	1	i	−i	i	−i
$\left[\frac{\pi}{2}, \pi\right]$	1	i	−i	−1	−1	−i	i	−i	i
$\left[\frac{\pi}{2}, \frac{\pi}{2}\right]$	1	i	−i	i	−i	−1	−1	1	1

coefficients $\hat{S}[n_{12}, n_{34}](T)$ can be found. In particular, we may extract $\hat{S}[-1, 1](T)$ by weighting each of the signals in the following combination,

$$\begin{aligned}\hat{S}[-1,1](T) = \frac{1}{8}\Bigg(& (1+i)S_{PP'}[0,0](T) + (-1+i)S_{PP'}[0,\pi](T) + (-2i)S_{PP'}\left[0,\frac{\pi}{2}\right](T) \\ & + (-i)S_{PP'}\left[-\frac{\pi}{2},0\right](T) + (1)S_{PP'}\left[-\frac{\pi}{2},\pi\right](T) \\ & + (-1+i)S_{PP'}\left[-\frac{\pi}{2},\frac{\pi}{2}\right](T) + (-1)S_{PP'}\left[\frac{\pi}{2},0\right](T) \\ & + (-i)S_{PP'}\left[\frac{\pi}{2},\pi\right](T) + (1+i)S_{PP'}\left[\frac{\pi}{2},\frac{\pi}{2}\right](T)\Bigg).\end{aligned} \tag{5.25}$$

As usual, we may dissect $\hat{S}[-1, 1](T)$ in terms of its ESA, SE and GSB components,

$$\hat{S}[-1,1](T) = \hat{S}_{\mathrm{ESA}}[-1,1](T) + \hat{S}_{\mathrm{SE}}[-1,1](T) + \hat{S}_{\mathrm{GSB}}[-1,1](T). \tag{5.26}$$

By selecting the terms in equations (5.21)–(5.23) that are proportional to $e^{i(-\phi_{12}+\phi_{34})}$, and developing the overlaps in analogy to equations (5.4)–(5.6), each of the (complex-valued) contributions to $\hat{S}[-1, 1]$ reads,

$$\begin{aligned}\hat{S}_{\mathrm{ESA}}[1,1](T) &= \langle \Psi_{14} | \mathbb{P}_{\mathrm{DEM}} | \Psi_{23} \rangle \\ &= \sum_{ijpq} \Omega^{3}_{fi} \Omega^{2}_{qg} \Omega^{\bar{1}}_{gp} \Omega^{\bar{4}}_{jf} \chi_{ijqp}(T),\end{aligned} \tag{5.27a}$$

Table 5.3. Different preparations and detections for the $\hat{S}_{PP'}[-1, 1](T)$ signal.

Possible initial states			Possible final states		
Pulse 1	Pulse 2	$\|q\rangle\langle p\|$	Pulse 3	Pulse 4	$\|i\rangle\langle j\|$
−	−	$\|\alpha\rangle\langle\alpha\|$	−	−	$\|\alpha\rangle\langle\alpha\|$ and $\|\beta\rangle\langle\beta\|$
−	+	$\|\beta\rangle\langle\alpha\|$	−	+	$\|\beta\rangle\langle\alpha\|$
+	−	$\|\alpha\rangle\langle\beta\|$	+	−	$\|\alpha\rangle\langle\beta\|$
+	+	$\|\beta\rangle\langle\beta\|$	+	+	$\|\alpha\rangle\langle\alpha\|$ and $\|\beta\rangle\langle\beta\|$

$$\hat{S}_{\rm SE}[1, 1](T) = -\langle\Psi_{13}|\mathbb{P}_{\rm GSM}|\Psi_{24}\rangle = -\sum_{ijpq}\Omega^{\bar{4}}_{gi}\Omega^{2}_{qg}\Omega^{\bar{1}}_{gp}\Omega^{3}_{jg}\chi_{ijqp}(T), \tag{5.27b}$$

$$\hat{S}_{\rm GSB}[1, 1](T) = -\langle\Psi_4|\Phi_{213}\rangle - \langle\Phi_{124}|\Psi_3\rangle = -\sum_{ip}\Omega^{3}_{ig}\Omega^{\bar{1}}_{gp}\Omega^{2}_{pg}\Omega^{\bar{4}}_{gi}. \tag{5.27c}$$

What have we accomplished? Through PC, we have isolated a quadrilinear signal, that is, an $O(\eta^4)$ signal that is $O(\eta)$ with respect to each of the pulses [9]. Now, each of the pulses can be tuned in the frequency domain. By using the two-color pulse toolbox of equation (5.12) for each of the four pulses (instead of just two), we obtain 16 $\hat{S}[-1, 1](T)$ signals where specific Liouville pathways are turned on and off. Linear inversion of $\chi(T)$ from these data is simple if the dipoles and the amplitudes of the fields are known.[2] The two pump pulses determine the initial state $|q\rangle\langle p|$, which then evolves until the two probe pulses detect the resulting final state $|i\rangle\langle j|$ at time T; see table 5.3.

As shown in example 1, part 4-d, for a two-level system in the SEM, there are 12 real-valued linearly independent elements of $\chi(T)$. Table 5.4 enumerates 12 spectroscopic signals (counting real and imaginary parts as separate signals) using seven experimental configurations, from which the elements of $\chi(T)$ can be extracted. Once Hermiticity and trace-preservation constraints (equations (1.16)–(1.19)) are taken into account, each of the real and imaginary parts of the $\hat{S}[-1, 1](T)$ signal reports on a single element of $\chi(T)$. For instance, although the first experiment $\{-,-,-,-\}$ reports on both $\chi_{\alpha\alpha\alpha\alpha}(T)$ and $\chi_{\beta\beta\alpha\alpha}(T)$, one of the variables can be eliminated using the trace preservation condition (equation (5.16)). For signals associated with complex-valued elements of $\chi(T)$, their real and imaginary parts report on the real and imaginary parts of the corresponding entry of $\chi(T)$, respectively. For instance, the real part of $\hat{S}[-1, 1](T)$ in the $\{+,-,+,-\}$ configuration is proportional to $\Re\chi_{\alpha\beta\alpha\beta}(T) = \Re\chi_{\beta\alpha\beta\alpha}(T)$, whereas the imaginary part is proportional to $\Im\chi_{\alpha\beta\alpha\beta}(T) = -\Im\chi_{\beta\alpha\beta\alpha}(T)$.

Physically, rows 1 and 4 of table 5.4 correspond to population transfer, 2 and 3 to regular coherence and decoherence dynamics, 5–8 to population-to-coherence transfer, 9 and 10 to coherence-to-population transfer, and 11 and 12 to coherence-to-coherence

[2] By exploiting properties of $\chi(T)$, it is possible to obtain the elements of $\chi(T)$ from the PP' data even without robust knowledge of each of the transition dipoles, i.e., $\chi_{abcd}(0) = \delta_{ac}\delta_{bd}$. Hence, if a given signal is proportional to a single element of $\chi(T)$, fixing the initial condition circumvents the need to know the values of the dipoles.

Table 5.4. A possible list of $\hat{S}_{PP'}[-1, 1](T)$ signals that suffice to extract all the linearly independent elements of $\chi(T)$.

1	2	3	4	$\lvert i\rangle\langle j\rvert$
−	−	−	−	$\chi_{\alpha\alpha\alpha\alpha}(T) = 1-\chi_{\beta\beta\alpha\alpha}(T)$
+	−	+	−	$\Re\chi_{\alpha\beta\alpha\beta}(T) = \Re\chi_{\beta\alpha\beta\alpha}(T)$,
				$\Im\chi_{\alpha\beta\alpha\beta}(T) = -\Im\chi_{\beta\alpha\beta\alpha}(T)$
+	+	+	+	$\chi_{\beta\beta\beta\beta}(T) = 1-\chi_{\alpha\alpha\beta\beta}(T)$
−	−	+	−	$\Re\chi_{\alpha\beta\alpha\alpha}(T) = \Re\chi_{\beta\alpha\alpha\alpha}(T)$,
				$\Im\chi_{\alpha\beta\alpha\alpha}(T) = -\Im\chi_{\beta\alpha\alpha\alpha}(T)$,
+	+	+	−	$\Re\chi_{\alpha\beta\beta\beta}(T) = \Re\chi_{\beta\alpha\beta\beta}(T)$,
				$\Im\chi_{\alpha\beta\beta\beta}(T) = -\Im\chi_{\beta\alpha\beta\beta}(T)$
+	−	−	−	$\Re\chi_{\alpha\alpha\alpha\beta}(T) = \Re\chi_{\alpha\alpha\beta\alpha}(T) = -\Re\chi_{\beta\beta\alpha\beta}(T) = -\Re\chi_{\beta\beta\beta\alpha}(T)$,
				$\Im\chi_{\alpha\alpha\alpha\beta}(T) = -\Im\chi_{\alpha\alpha\beta\alpha}(T) = -\Im\chi_{\beta\beta\alpha\beta}(T) = \Im\chi_{\beta\beta\beta\alpha}(T)$
+	−	−	+	$\Re\chi_{\beta\alpha\alpha\beta}(T) = \Re\chi_{\alpha\beta\beta\alpha}(T)$
				$\Im\chi_{\beta\alpha\alpha\beta}(T) = -\Im\chi_{\alpha\beta\beta\alpha}(T)$

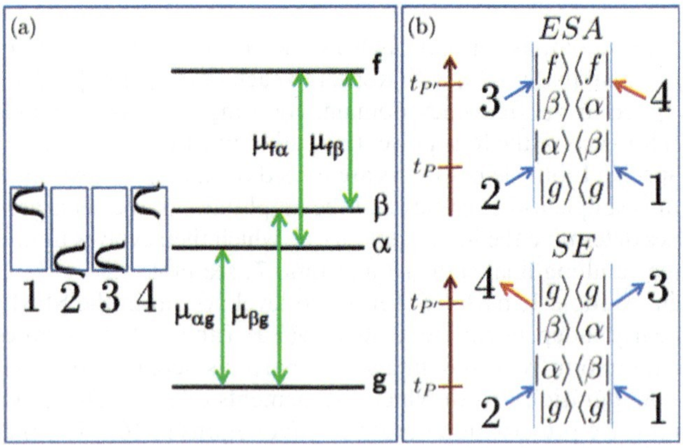

Figure 5.6. Example of a sequence to detect a particular element of $\chi(T)$, namely $\chi_{\beta\alpha\alpha\beta}(T)$, corresponding to rows 11 and 12 of table 5.4. By selectively tuning pulses 1, 2, 3 and 4 to be $\{+,-,-,+\}$ (resonant with the $g \leftrightarrow \beta$, $g \leftrightarrow \alpha$, $g \leftrightarrow \alpha$, $g \leftrightarrow \beta$ transitions, respectively), we are able to obtain a signal that depends on a single $\chi(T)$ element, namely $\chi_{\beta\alpha\alpha\beta}(T)$. (*a*) Approximate energy level diagram with a schematic of transitions and pulses. Note that, in a real experiment, the states are broadened due to coupling of the electronic system $\mathscr{S}$ to the bath $\mathscr{B}$. By tuning the value of the relative phase γ between pulses (not shown in the figure), we can access the real and the imaginary parts of $\chi_{\beta\alpha\alpha\beta}(T)$. (*b*) DS-FDs associated with the detection of $\chi_{\beta\alpha\alpha\beta}(T)$.

transfer. Figure 5.6 summarizes the idea for extracting the element $\chi_{\beta\alpha\alpha\beta}(T)$. Note that the experiments corresponding to population transfer dynamics can be carried out with the standard two-pulse PP' setup, as described in table 5.1.

We note that the T-independent background due to $S_{\mathrm{GSB}}(T)$ vanishes for settings where the first and the second pulses are different, since the action of the those two simultaneous pulses on the initial state $|g\rangle\langle g|$ cannot bring it back to $|g\rangle\langle g|$.

We mention that similar studies of quantum state reconstruction in the context of chemical physics have been reported before [1–6, 9, 10, 12, 17]. In most of these studies, the goal is to reconstruct *nuclear* wavepackets in various PES for specific pathways of excitation and de-excitation. In our work, we aim for full QPT for the reduced electronic state of the multichromophoric system, where we are not interested in determing the particular nuclear states involved.

Example 10. Transient grating spectroscopy and QPT

In a variation on the previous example, we now show that transient grating (TG) spectroscopy can also be used for QPT. It is also a four-wave mixing technique, except that the wavevectors $\mathbf{k}_1$, $\mathbf{k}_2$, $\mathbf{k}_3$, $\mathbf{k}_4$, are mutually noncollinear, yet obey the phase-matching constraint $\mathbf{k}_4 = \mathbf{k}_{TG} \equiv -\mathbf{k}_1 + \mathbf{k}_2 + \mathbf{k}_3$. Furthermore, the pulses arrive at the sample in pairs, $t_1 = t_2 = t_P$ and $t_3 = t_4 = t_{P'}$. We define the TG signal S_{TG} (in analogy to $S_{PP'}$, equation (4.18)) as [11, 19]

$$\begin{aligned} S_{TG}[\gamma] &= S_4\ (\text{with } 1, 2, 3) - S_4\ (\text{without } 1, 2, 3) \\ &= 2\Im \int_{-\infty}^{\infty} dt' \varepsilon_4^*(t' - t_4) e^{-i\phi_4} \mathbf{e}_{4'}^* \cdot (\mathbf{P}_{\mathbf{k}_{TG}}(t') - \mathbf{P}^{(1)}_{\mathbf{k}_{TG}}) e^{i(-\phi_1+\phi_2+\phi_3)} \\ &= 2\Im \int_{-\infty}^{\infty} dt' \varepsilon_4^*(t' - t_{P'}) \mathbf{e}_4^* \cdot \mathbf{P}^{(3)}_{\mathbf{k}_{TG}}(t') e^{i\gamma}, \end{aligned} \tag{5.28}$$

where S_4 (with 1, 2, 3) is the number of photons lost by pulse 4 after passing through the sample, which had previously interacted with pulses 1, 2 and 3. S_4 (without 1, 2, 3) is the corresponding (linear) absorption of pulse 4 in the absence of the other pulses (see figure 5.7). Just as with the PP' signal, the lowest-order polarization that contributes to S_{TG} is $\mathbf{P}^{(3)}_{\mathbf{k}_{TG}}$. Also, the signal is seen to depend on the relative phase,

$$\gamma \equiv -\phi_{12} + \phi_{34}, \tag{5.29}$$

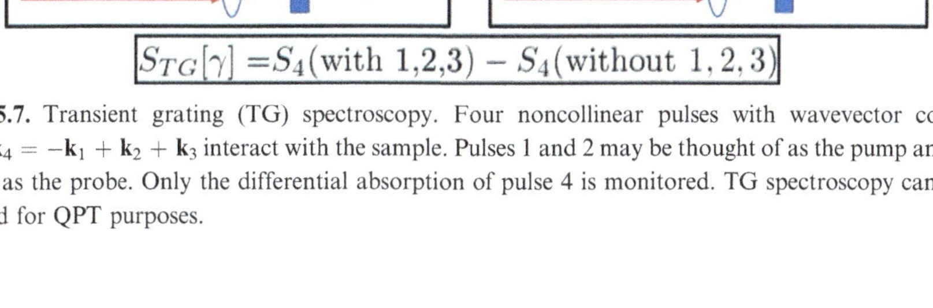

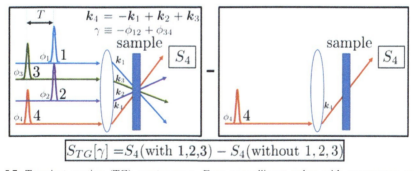

Figure 5.7. Transient grating (TG) spectroscopy. Four noncollinear pulses with wavevector constraint $\mathbf{k}_{TG} = \mathbf{k}_4 = -\mathbf{k}_1 + \mathbf{k}_2 + \mathbf{k}_3$ interact with the sample. Pulses 1 and 2 may be thought of as the pump and pulses 3 and 4 as the probe. Only the differential absorption of pulse 4 is monitored. TG spectroscopy can also be exploited for QPT purposes.

which is the already familiar phase combination from the previous example. As usual, we may decompose the TG signal as,

$$S_{\mathrm{TG}}[\gamma] = S_{\mathrm{ESA}}[\gamma] + S_{\mathrm{SE}}[\gamma] + S_{\mathrm{GSB}}[\gamma]. \tag{5.30}$$

1. Derive expressions for S_{ESA}, S_{SE} and S_{GSB} in terms of wavepacket overlaps, as in equations (4.27*a*), (4.28*a*) and (29*a*). To establish a concrete comparison with the previous example, define the closely related PC signal $S^{\mathrm{PC}}[\gamma]$,

$$\begin{aligned} S_{\mathrm{PC}}[\gamma] &\equiv \Re\{\hat{S}[-1, 1]\mathrm{e}^{\mathrm{i}\gamma}\} \\ &= S_{\mathrm{PC,ESA}}[\gamma] + S_{\mathrm{PC,SE}}[\gamma] + S_{\mathrm{PC,GSB}}[\gamma], \end{aligned} \tag{5.31}$$

where $\hat{S}[-1, 1]$ is the signal constructed in equations (5.25)–(5.27*c*), and,

$$S_{\mathrm{PC,ESA}}[\gamma] = \Re\{\langle\Psi_{14}|\mathbb{P}_{\mathrm{DEM}}|\Psi_{23}\rangle\mathrm{e}^{\mathrm{i}\gamma}\} \tag{5.32a}$$

$$= \Re\sum_{ijpq}\Omega^{3}_{fi}\Omega^{2}_{qg}\Omega^{\bar{1}}_{gp}\Omega^{\bar{4}}_{jf}\chi_{ijqp}(T), \tag{5.32b}$$

$$S_{\mathrm{PC,SE}}[\gamma] = -\Re\{\langle\Psi_{13}|\mathbb{P}_{\mathrm{GSM}}|\Psi_{24}\rangle\mathrm{e}^{\mathrm{i}\gamma}\} \tag{5.32c}$$

$$= -\Re\sum_{ijpq}\Omega^{\bar{4}}_{gi}\Omega^{2}_{qg}\Omega^{\bar{1}}_{gp}\Omega^{3}_{jg}\chi_{ijqp}(T), \tag{5.32d}$$

$$S_{\mathrm{PC,GSB}}[\gamma] = -\Re\{(\langle\Psi_{4}|\Phi_{213}\rangle + \langle\Phi_{124}|\Psi_{3}\rangle)\mathrm{e}^{\mathrm{i}\gamma}\} \tag{5.32e}$$

$$= -\Re\sum_{ip}\Omega^{3}_{ig}\Omega^{\bar{1}}_{gp}\Omega^{2}_{pg}\Omega^{\bar{4}}_{gi}. \tag{5.32f}$$

Clearly, $S_{\mathrm{PC}}[\gamma]$ is just another form to write $\hat{S}[-1, 1]$; explicitly, $\hat{S}[-1, 1] = S_{\mathrm{PC}}[0] + \mathrm{i}S_{\mathrm{PC}}[-\frac{\pi}{2}]$. As we showed in the last example, QPT can be carried out with the information of $\hat{S}[-1, 1]$. Therefore, it can also be performed with $S_{\mathrm{PC}}[\gamma]$.

2. Evaluate $S_{\mathrm{ESA}}[\gamma]$, $S_{\mathrm{SE}}[\gamma]$, and $S_{\mathrm{GSB}}[\gamma]$ for the vibrationless coupled dimer from section 3.2.
3. It is instructive to think about $S_{\mathrm{TG}}[\gamma]$ as a generalization of the PP' signal $S_{PP'}$. Under what limits can $S_{\mathrm{TG}}[\gamma]$ be exactly reduced to $S_{PP'}$?
4. It is also customary to present the TG signal as a complex-valued signal,

$$\Sigma_{\mathrm{TG}}(T) \equiv -i\int_{-\infty}^{\infty}\mathrm{d}t'\varepsilon_4^*(t' - t_{P'})\mathbf{e}_4^*\cdot\mathbf{P}^{(3)}_{\mathbf{k}_{\mathrm{TG}}}(t'). \tag{5.33}$$

Comparing equations (5.28) and (5.33), what combination of $S_{\mathrm{TG}}[\gamma]$ signals does one need to measure to obtain $\Sigma_{\mathrm{TG}}(T)$?

Solution

1. In analogy to equation (3.33), the third-order polarization along the $\mathbf{k}_{\mathrm{TG}} = -\mathbf{k}_1 + \mathbf{k}_2 + \mathbf{k}_3$ direction is given by,

$$\begin{aligned} \mathbf{P}^{(3)}_{\mathbf{k}_{\mathrm{TG}}}(t) = &\langle\Psi_{+1}(t)|\boldsymbol{\mu}|\Psi_{+2+3}(t)\rangle + \langle\Psi_{+1-3}(t)|\boldsymbol{\mu}|\Psi_{+2}(t)\rangle \\ &+ \langle\Psi_{+1-2}(t)|\boldsymbol{\mu}|\Psi_{+3}(t)\rangle + \langle\Psi_{0}(t)|\boldsymbol{\mu}|\Psi_{+2-1+3}(t)\rangle \end{aligned} \tag{5.34}$$

Proceeding as in equations (4.34)–(4.36),

$$S_{\mathrm{ESA}}[\gamma] = 2\Im \int_{-\infty}^{\infty} \mathrm{d}t' \varepsilon_4^*(t' - t_4) \langle \Psi_{+1}(t') | \boldsymbol{\mu} \cdot \mathbf{e}_4^* | \Psi_{+2+3}(t') \rangle \mathrm{e}^{\mathrm{i}\gamma} \tag{5.35}$$

$$= 2\Re \int_{-\infty}^{\infty} \mathrm{d}t' [\partial_{t'} \langle \Psi_{14}(t') |] \mathbb{P}_{\mathrm{DEM}} | \Psi_{23}(t') \rangle \mathrm{e}^{\mathrm{i}\gamma}, \tag{5.36}$$

$$S_{\mathrm{SE}}[\gamma] \equiv 2\Im \int_{-\infty}^{\infty} \mathrm{d}t' \varepsilon_4^*(t' - t_4) \langle \Psi_{+1-3}(t') | \boldsymbol{\mu} \cdot \mathbf{e}_4^* | \Psi_{+2}(t') \rangle \mathrm{e}^{\mathrm{i}\gamma} \tag{5.37}$$

$$= 2\Re \int_{-\infty}^{\infty} \mathrm{d}t' [\partial_{t'} \langle \Psi_{13}(t') |] \mathbb{P}_{\mathrm{GSM}} | \Psi_{24}(t') \rangle \mathrm{e}^{\mathrm{i}\gamma}, \tag{5.38}$$

$$S_{\mathrm{GSB}}[\gamma] = 2\Im \int_{-\infty}^{\infty} \mathrm{d}t' \varepsilon_4^*(t' - t_{P'})$$

$$\times \left[\langle \Psi_{+1-2}(t') | \boldsymbol{\mu} \cdot \mathbf{e}_4^* | \Psi_{+3}(t) \rangle + \langle \Psi_0(t) | \boldsymbol{\mu} \cdot \mathbf{e}_4^* | \Psi_{+2-1+3}(t') \rangle \right] \mathrm{e}^{\mathrm{i}\gamma} \tag{5.39}$$

$$= 2\Re \int_{-\infty}^{\infty} \mathrm{d}t' [\partial_{t'} \langle \Phi_{124}(t') |] | \Psi_3(t') \rangle + [\partial_{t'} \langle \Psi_4(t') |] | \Phi_{213}(t') \rangle \mathrm{e}^{\mathrm{i}\gamma}. \tag{5.40}$$

Although it may appear that the integrals are cancelled by the derivatives in the integrands, this is not possible in general. For example, the overlap $\langle \Psi_{14} | \mathbb{P}_{\mathrm{DEM}} | \Psi_{23} \rangle$ in $S_{\mathrm{ESA}}^{\mathrm{PC}}[\gamma]$, equation (5.32*a*), satisfies the identity,

$$\langle \Psi_{14} | \mathbb{P}_{\mathrm{DEM}} | \Psi_{23} \rangle = \int_{-\infty}^{\infty} \mathrm{d}t' \partial_{t'} [\langle \Psi_{14}(t') | \mathbb{P}_{\mathrm{DEM}} | \Psi_{23}(t') \rangle]$$

$$= \int_{-\infty}^{\infty} \mathrm{d}t' [\partial_{t'} \langle \Psi_{14}(t') |] \mathbb{P}_{\mathrm{DEM}} | \Psi_{23}(t') \rangle$$

$$+ \int_{-\infty}^{\infty} \mathrm{d}t' \langle \Psi_{14}(t') | \mathbb{P}_{\mathrm{DEM}} [\partial_{t'} | \Psi_{23}(t') \rangle]. \tag{5.41}$$

Because there are two terms in equation (5.41), only one of which corresponds to equation (5.36), $S_{\mathrm{ESA}}[\gamma] \neq S_{\mathrm{ESA}}^{\mathrm{PC}}[\gamma]$ in general. This issue did not arise in the derivation of the expression for linear absorption in equation (4.5) or of the equations for PP' spectroscopy in (4.34)–(4.36). This is because the 'missing' term $2\Re\{\int_{-\infty}^{\infty} \mathrm{d}t' \langle \Psi_{14}(t') | \mathbb{P}_{\mathrm{DEM}} (\partial_{t'} | \Psi_{23}(t') \rangle) \mathrm{e}^{\mathrm{i}\gamma}\}$ corresponds to the ESA signal along the direction $\mathbf{k}_3 = -\mathbf{k}_2 + \mathbf{k}_1 + \mathbf{k}_4$. Physically, the energy from both pulses 3 and 4, and not just 4, is involved in creating the DEM overlap $\langle \Psi_{14} | \mathbb{P}_{\mathrm{DEM}} | \Psi_{23} \rangle$. Interestingly, averaging the TG signals from both $\mathbf{k}_3$ and $\mathbf{k}_4$ directions yields precisely $S^{\mathrm{PC}}[\gamma]$, which is not surprising if we think about pulses 1 and 2 as the pump and pulses 3 and 4 as the probe.

2. Even though the TG signal alone cannot be readily interpreted in terms of simple wavepacket overlaps, its evaluation is not too difficult.

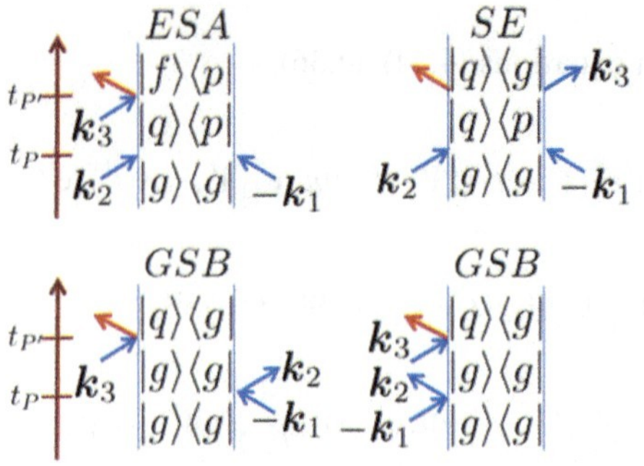

Figure 5.8. DS-FDs for the TG nonlinear polarization component $\mathbf{P}^{(3)}_{\mathbf{k}_{TG}}(t)$. Notice similarities with figure 3.2(*b*) for $\mathbf{P}^{(3)}_{\mathbf{k}_{P'}}$ in the context of *PP'* spectroscopy.

First, we calculate S_{ESA} for the vibrationless coupled dimer (see section 3.2) by evaluating equation (5.36) explicitly (see DS-FDs in figure 5.8). This calculation is similar to the one performed in example 6. The wavefunctions of interest are

$$|\Psi_{+1}(t)\rangle = \mathrm{i}\int_0^t \mathrm{d}t' \mathrm{e}^{-\mathrm{i}\omega_q(t-t')}[\boldsymbol{\mu}\cdot\mathbf{e}_1\varepsilon_1(t'-t_P)]\mathrm{e}^{-\mathrm{i}\omega_g t'}|g\rangle$$
$$\approx \mathrm{i}\sum_{q=\alpha,\beta}|q\rangle \mathrm{e}^{-\mathrm{i}\omega_q(t-t_P)}\Omega^1_{qg}\mathrm{e}^{-\mathrm{i}\omega_g t_P}, \quad (5.42)$$

where we have used equation (3.19*a*) for $n = 1$, $\int_0^t \mathrm{d}t' \approx \int_{-\infty}^t \mathrm{d}t'$ and

$$|\Psi_{+2+3}(t)\rangle = (-\mathrm{i})^2\int_0^t \mathrm{d}t'\int_0^{t'} \mathrm{d}t'' \mathrm{e}^{-\mathrm{i}H_0(t-t')}[-\boldsymbol{\mu}\cdot\mathbf{e}_{P'}\varepsilon_3(t'-t_{P'})]$$
$$\times \mathrm{e}^{-\mathrm{i}H_0(t'-t'')}\{-\boldsymbol{\mu}\cdot\mathbf{e}_2\varepsilon_2(t''-t_P)\}\mathrm{e}^{-\mathrm{i}H_0t''}|g\rangle$$
$$\approx -|f\rangle\sum_{q=\alpha,\beta}\mathrm{e}^{-\mathrm{i}\omega_f(t-t_{P'})}\int_{-\infty}^t \mathrm{d}t'' \mathrm{e}^{\mathrm{i}\omega_{fq}(t''-t_P)}\varepsilon_3(t''-t_{P'})$$
$$\times(\boldsymbol{\mu}_{fq}\cdot\mathbf{e}_3)\mathrm{e}^{-\mathrm{i}\omega_q T}\Omega^2_{qg}\mathrm{e}^{-\mathrm{i}\omega_g t_P}, \quad (5.43)$$

which is the analogue of equation (4.45). Here, we have approximated $\int_0^\infty \mathrm{d}t'' \approx \int_{-\infty}^\infty \mathrm{d}t''$ and $\int_0^t \mathrm{d}t' \approx \int_{-\infty}^t \mathrm{d}t'$. Altogether,

$$\langle\Psi_{+1}(t')|\boldsymbol{\mu}\cdot\mathbf{e}_4^*|\Psi_{+2+3}(t')\rangle = \mathrm{i}\sum_{q,p=\alpha,\beta}\int_{-\infty}^{t'} \mathrm{d}t'' \mathrm{e}^{\mathrm{i}\omega_{fq}(t''-t_P)}\varepsilon_3(t''-t_{P'})$$
$$\times(\boldsymbol{\mu}_{fq}\cdot\mathbf{e}_3)\Omega^2_{qg}\Omega^{\bar{1}}_{gp}(\boldsymbol{\mu}_{pf}\cdot\mathbf{e}_4^*)\mathrm{e}^{-\mathrm{i}\omega_{fp}(t'-t_{P'})}\mathrm{e}^{-\mathrm{i}\omega_{qp}T}. \quad (5.44)$$

Finally, using the nested integral introduced in chapter 3, footnote 5 (page 3–10), we obtain,

$$\int_{-\infty}^{\infty} dt' \varepsilon_4^*(t' - t_{P'}) e^{-i\omega_{fp}(t'-t_{P'})} \int_{-\infty}^{t'} dt'' \varepsilon_3(t'' - t_{P'}) e^{i\omega_{fq}(t''-t_{P'})}$$
$$= \frac{\zeta_{fq,pf}^{3,\bar{4}}}{2} \tilde{\varepsilon}_3(\omega_{fq}) \tilde{\varepsilon}_4^*(\omega_{fp}), \tag{5.45}$$

where

$$\zeta_{fq,pf}^{3,\bar{4}} \equiv 1 - \operatorname{erf}\left(\frac{i(\omega_{fq} - \omega_3 + \omega_{fp} - \omega_4)\sigma}{2}\right). \tag{5.46}$$

Equation (5.45) would be the product $\tilde{\varepsilon}_3(\omega_{fq})\tilde{\varepsilon}_4^*(\omega_{fp})$ of Fourier transforms if the two integrals were independent. Since they are not, the result includes the interference term between pulses 3 and 4, $\zeta_{fq,pf}^{3,\bar{4}}/2$. Substituting equations (5.44) and (5.45) into equation (5.35) yields,

$$S_{\mathrm{ESA}}[\gamma](T) = 2\Im \int_{-\infty}^{\infty} dt' \varepsilon_4^*(t' - t_4) \langle \Psi_{+1}(t') | \boldsymbol{\mu} \cdot \mathbf{e}_4^* | \Psi_{+2+3}(t') \rangle e^{i\gamma}$$
$$= \Re \sum_{p,q=\alpha,\beta} \zeta_{fq,pf}^{3,\bar{4}} \Omega_{fq}^{3} \Omega_{qg}^{2} \Omega_{gp}^{\bar{1}} \Omega_{pf}^{\bar{4}} e^{i\gamma} e^{-i\omega_{qp}T}. \tag{5.47}$$

Analogous steps yield

$$S_{\mathrm{SE}}[\gamma](T) = \Re \sum_{p,q=\alpha,\beta} \zeta_{pg,gp}^{3,\bar{4}} \Omega_{gq}^{\bar{4}} \Omega_{qg}^{2} \Omega_{gp}^{\bar{1}} \Omega_{pg}^{3} e^{i\gamma} e^{-i\omega_{qp}T}, \tag{5.48}$$

$$S_{\mathrm{GSB}}[\gamma](T) = \Re \sum_{p,q=\alpha,\beta} \left[\frac{\zeta_{gp,pg}^{\bar{1},2} + \left(\zeta_{gp,pg}^{\bar{1},2}\right)^*}{2}\right] \zeta_{qg,gq}^{3,\bar{4}} \Omega_{gq}^{\bar{4}} \Omega_{qg}^{3} \Omega_{gp}^{\bar{1}} \Omega_{pg}^{2} e^{i\gamma}. \tag{5.49}$$

3. A comparison of equations (5.28) and (5.34) with equations (4.18) and (4.19) reveals that S_{TG} is identical to $S_{PP'}$ when pulses 1 and 2, and, independently, 3 and 4 coincide, in which case $\gamma = 0$.
4. Inspection of equations (5.28) and (5.33) immediately yields

$$\Sigma_{\mathrm{TG}}(T) = \frac{1}{2}\left\{S_{\mathrm{TG}}[0](T) + iS_{\mathrm{TG}}\left[-\frac{\pi}{2}\right](T)\right\}, \tag{5.50}$$

which means that the complex-valued TG signal may be constructed using two independent TG signals and choosing the phases to be $\gamma = 0, -\frac{\pi}{2}$.

Let us assume that the pulses have been chosen so that the sum of the carrier frequencies of pulses 3 and 4, $\omega_3 + \omega_4$, matches the sum of the respective transition energies $\omega_{fq} + \omega_{fp}$, the argument of the error function in equation (5.46) vanishes, and we have $\zeta_{fq,pf}^{3\bar{4}} = 1$. Suppose that this is also true for the other ζ functions, $\zeta_{pg,gp}^{3,\bar{4}} = \zeta_{qg,gq}^{3,\bar{4}} = \zeta_{gp,pg}^{\bar{1},2} = 1$.[3] In which case, $S_{\mathrm{TG}}[\gamma]$ *does* coincide with $S_{\mathrm{PC}}[\gamma]$, as can be

[3] Although this condition is not necessary for QPT, it simplifies the evaluation of transition amplitudes.

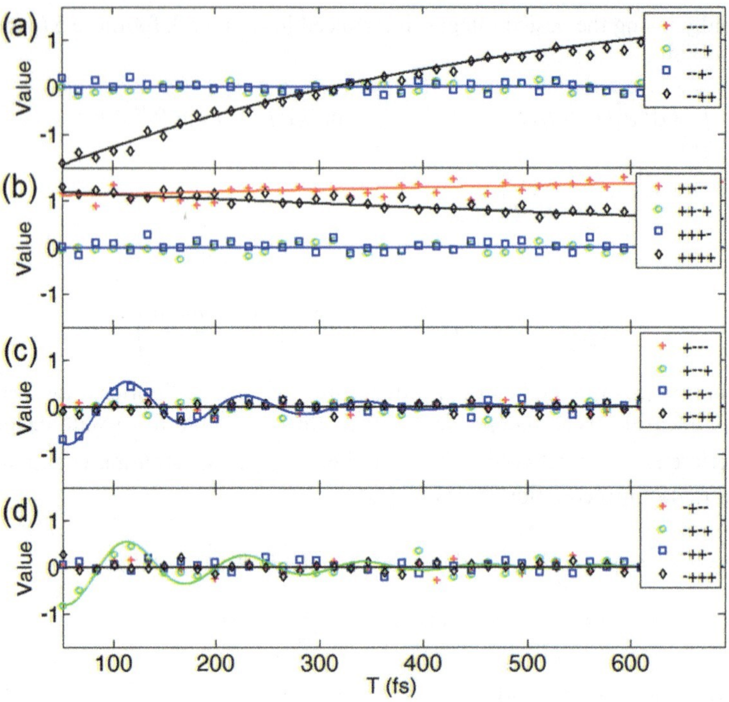

Figure 5.9. Simulated data for TG signals from 16 two-color experiments. The legends *pqrs* correspond to the real parts of the isotropically-averaged signals $S_{\text{TG}}(T)$, where the pulses are selectively resonant with the transitions $|g\rangle \leftrightarrow \{|p\rangle, |q\rangle, |r\rangle, |s\rangle\}$. This can be achieved using pulses of two different colors, such that pulse $-$ is selective to the $|g\rangle \leftrightarrow |\alpha\rangle$ transition, while pulse $+$ is selective to the $|g\rangle \leftrightarrow |\beta\rangle$ transition. The panels are organized by QPT initial state: (*a*) for $|\alpha\rangle\langle\alpha|$, (*b*) for $|\beta\rangle\langle\beta|$, (*c*) for $|\alpha\rangle\langle\beta|$, and (*d*) for $|\beta\rangle\langle\alpha|$). The ideal signals are depicted as continuous functions, whereas the simulations with inhomogeneous broadening and noise are represented as discrete points of the same color. (Reprinted from Yuen-Zhou J, Krich J J, Mohseni M and Aspuru-Guzik A 2011 *Proc. Natl Acad. Sci. USA* **108**(43) 17615.)

verified by comparing equations (5.47)–(5.49) to equations (5.32*b*), (5.32*d*) and (5.32*f*), and setting $\chi_{ijqp}(T) = \delta_{iq}\delta_{jp}\mathrm{e}^{-\mathrm{i}\omega_{qp}T}$. Since $S_{\text{PC}}[\gamma]$ can be used to perform full QPT on the SEM of a coupled dimer (example 9), so can $S_{\text{TG}}[\gamma]$. Figure 5.9 shows simulations of noisy TG signals for a coupled dimer under 16 different pulse configurations, and fixed at $\gamma = 0$ [23]. In these simulations, each pulse is chosen to be selectively resonant either with the $|g\rangle \leftrightarrow |\alpha\rangle$, $|\beta\rangle \leftrightarrow |f\rangle$ (pulse $-$) or the $|g\rangle \leftrightarrow |\beta\rangle$, $|\alpha\rangle \leftrightarrow |f\rangle$ (pulse $+$) transitions. The $\gamma = -\frac{\pi}{2}$ traces for the imaginary parts of $\chi(T)$ (not shown) were also computed, yielding a total of 32 time traces. In principle, only 12 time traces are necessary, but here we have used all the available information to yield a robust inversion of the SEM dynamics.

A straightforward inversion of $\chi(T)$ from the time traces is often ill-behaved. To ensure that $\chi(T)$ is Hermitian positive-semidefinite (equation (1.20)), $\chi(T)$ is inverted via a semi-definite programming procedure [7, 8]. Results obtained in this way are shown in figure 5.10 [23]. This demonstrates a full simulation of QPT from noisy experimental data using the TG method described here.

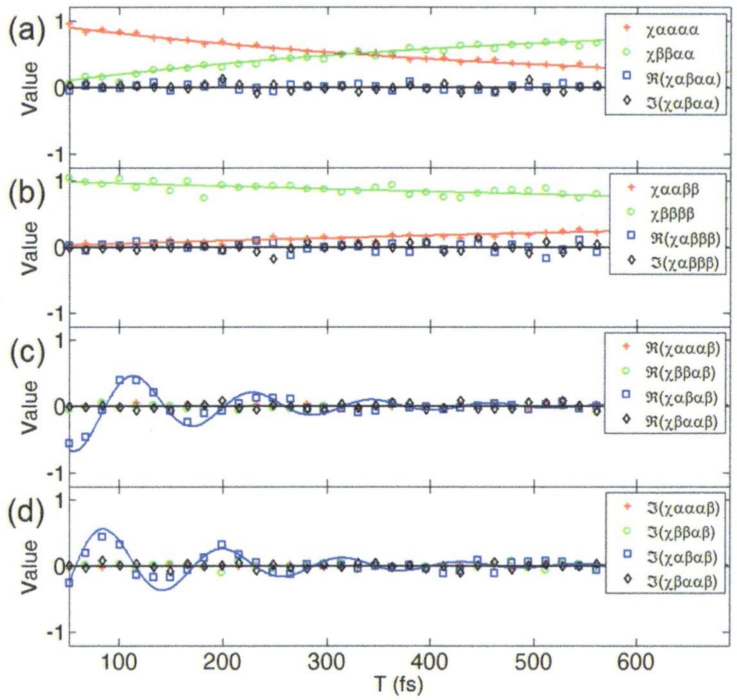

Figure 5.10. Elements of $\chi(T)$ extracted by inverting the TG signals in figure 5.9. The ideal values are shown as a continuous function, whereas the discrete points are those extracted from noisy data. The panels are organized according to the initial state, (*a*) for $|\alpha\rangle\langle\alpha|$, (*b*) for $|\beta\rangle\langle\beta|$ and (*c*) and (*d*) for $|\alpha\rangle\langle\beta|$. (Reprinted from Yuen-Zhou J, Krich J J, Mohseni M and Aspuru-Guzik A 2011 *Proc. Natl Acad. Sci. USA* **108**(43) 17615.)

We have now completed our journey from the beginning of ultrafast optical spectroscopy and wavepackets to demonstrate the method of QPT for the SEM of model systems. Since QPT gives in principle the maximum information possible about a quantum system, such techniques should prove useful in characterizing multichromophoric systems.

Bibliography

[1] Avisar D and Tannor D J 2011 Complete reconstruction of the wave function of a reacting molecule by four-wave mixing spectroscopy *Phys. Rev. Lett.* **106** 170405

[2] Avisar D and Tannor D J 2011 Wavepacket and potential reconstruction by four-wave mixing spectroscopy: preliminary application to polyatomic molecules *Faraday Discuss.* **153** 131–48

[3] Avisar D and Tannor D J 2012 Multi-dimensional wavepacket and potential reconstruction by resonant coherent anti-Stokes Raman scattering: Application to H_2O and HOD *J. Chem. Phys.* **136** 214107

[4] Cina J A 2000 Nonlinear wavepacket interferometry for polyatomic molecules *J. Chem. Phys.* **113** 9488–96

[5] Cina J A 2008 Wave-packet interferometry and molecular state reconstruction: spectroscopic adventures on the left-hand side of the Schrödinger equation *Annu. Rev. Phys. Chem.* **59** 319–42

[6] Cina J A, Kilin D S and Humble T S 2003 Wavepacket interferometry for short-time electronic energy transfer: multidimensional optical spectroscopy in the time domain *J. Chem. Phys.* **118** 46–61

[7] Grant M and Boyd S 2008 *Graph Implementations for Nonsmooth Convex Programs* ed V Blondel, S Boyd and H Kimura *Recent Advances in Learning and Control* (*Lecture Notes in Control and Information Sciences* (Berlin: Springer)) pp 95–110

[8] Grant M and Boyd S 2011 Cvx: Matlab software for disciplined convex programming, version 1.21, April

[9] Humble T S and Cina J A 2004 Molecular state reconstruction by nonlinear wave packet interferometry *Phys. Rev. Lett.* **93** 060402

[10] Humble T S and Cina J A 2006 Nonlinear wave-packet interferometry and molecular state reconstruction in a vibrating and rotating diatomic molecule *J. Phys. Chem.* B **110** 18879–92

[11] Kwak K, Minhaeng C, Fleming G R, Agarwal R and Prall B S 2003 Two-color transient grating spectroscopy of a two-level system *Bull. Korean Chem. Soc.* **24** 1069–74

[12] Leichtle C, Schleich W P, Averbukh I Sh and Shapiro M 1998 Quantum state holography *Phys. Rev. Lett.* **80** 1418–21

[13] Liu Z F, Yan H C, Wang K B, Kuang T Y, Zhang J P, Gui L L, An X M and Chang W R 2004 Crystal structure of spinach major light-harvesting complex at 2.72 angstrom resolution *Nature* **428** 287–92

[14] Lott G A, Perdomo-Ortiz A, Utterback J K, Widom J R, Aspuru-Guzik A and Marcus A H 2011 Conformation of self-assembled porphyrin dimers in liposome vesicles by phase-modulation 2d fluorescence spectroscopy *Proc. Natl Acad. Sci. USA* **108** 16521–6

[15] Rebentrost P, Shim S, Yuen-Zhou J and Aspuru-Guzik A 2011 Characterization and quantification of the role of coherence in ultrafast quantum biological experiments using quantum master equations, atomistic simulations, and quantum process tomography *Proc. Chem.* **3** 332–46

[16] Tan H S 2008 Theory and phase-cycling scheme selection principles of collinear phase coherent multi-dimensional optical spectroscopy *J. Chem. Phys.* **129** 124501

[17] Tekavec P F, Dyke T R and Marcus A H 2006 Wave packet interferometry and quantum state reconstruction by acousto-optic phase modulation *J. Chem. Phys.* **125** 194303

[18] Tekavec P F, Lott G A and Marcus A H 2007 Fluorescence-detected two-dimensional electronic coherence spectroscopy by acousto-optic phase modulation *J. Chem. Phys.* **127** 214307

[19] Womick J M and Moran A M 2009 Exciton coherence and energy transport in the light-harvesting dimers of allophycocyanin *J. Phys. Chem.* B **113** 15747–59

[20] Yan S and Tan H S 2009 Phase cycling schemes for two-dimensional optical spectroscopy with a pump–probe beam geometry *Chem. Phys.* **360** 110–5

[21] Yuen-Zhou J and Aspuru-Guzik A 2011 Quantum process tomography of excitonic dimers from two-dimensional electronic spectroscopy. I. General theory and application to homodimers *J. Chem. Phys.* **134** 134505

[22] Yuen-Zhou J, Krich J J and Aspuru-Guzik A 2012 A witness for coherent electronic vs vibronic-only oscillations in ultrafast spectroscopy *J. Chem. Phys.* **136** 234501

[23] Yuen-Zhou J, Krich J J, Mohseni M and Aspuru-Guzik A 2011 Quantum state and process tomography of energy transfer systems via ultrafast spectroscopy *Proc. Natl Acad. Sci. USA* **108** 17615–20

Chapter 6

Computational methods for spectroscopy simulations

So far, we have discussed how nonlinear spectroscopy can be understood, both using wavepackets and as an approach to quantum process tomography. We now turn to the problem of numerical simulation in this context. As noted before, wavepacket methods offer a manifestly time-dependent approach to ultrafast spectroscopy, and this strength should be reflected in any numerical methods used. Numerical simulations generally proceed by representing the nuclear wavepackets on a discrete grid and modeling its evolution through the use of a propagator, or alternatively, using swarms of semiclassical trajectories that when weighted appropriately reconstruct the wavepackets of interest. For clarity and brevity, we will only discuss the former, but we invite the reader to explore the literature on the latter [5, 7, 11].

In this chapter, we first describe the propagation of nuclear wavepackets in the absence of optical pulses (section 6.1) and show how to use this method to calculate linear absorption spectra (section 6.2). In section 6.3, we show how to incorporate optical pulses in order to simulate nonlinear spectra. We conclude in section 6.4 with some extensions of the method. This discussion is supplemented by example MATLAB® code provided online on the book's website.

6.1 Propagation of wavefunctions

A general molecular wavefunction consists of nuclear wavepackets in various PES. Setting aside the perturbative action of the pulses, we will first discuss how to evolve such a wavefunction using the propagator $U(\delta t) = e^{-iH_0(\mathbf{R})\delta t}$ for a time-independent molecular Hamiltonian $H_0(\mathbf{R})$ (equation (2.1)) and a time step δt. Clearly, having access to $U(\delta t)$ also gives access to the propagator at longer times by the composition property, $U(t) = [U(\delta t)]^{N_t}$, indicating successive applications of the same δt propagator for $N_t = t/\delta t$ steps. The free evolution of the wavefunction due to $H_0(\mathbf{R})$ can be carried out efficiently when $H_0(\mathbf{R})$ is block-diagonal, i.e., does not mix different excitation manifolds. Hence, as explained in previous sections, we do not treat situations that,

doi:10.1088/978-0-750-31062-8ch6

in the absence of light perturbations, involve substantial nonadiabatic dynamics between different excitation manifolds.

The propagator $U(\delta t)$ can be numerically implemented in numerous ways, including the discrete variable representation, the Chebyshev and the split-operator (SO) methods [3, 10, 14]. We will restrict ourselves to describing the SO method, which is used in the provided MATLAB® code.

We write $H_0(\mathbf{R}) = T_N + H_{el}(\mathbf{R})$, with T_N the nuclear kinetic energy and $H_{el}(\mathbf{R})$ the effective potential produced by the electrons at each nuclear coordinate R, including the effect of the nuclear–nuclear repulsion. In the SO method the propagator is approximated as,

$$\begin{aligned} U(\delta t) &= \mathrm{e}^{-\mathrm{i}(T_N + H_{el}(\mathbf{R}))\delta t} \\ &\approx \mathrm{e}^{-\mathrm{i}H_{el}(\mathbf{R})\frac{\delta t}{2}}\mathrm{e}^{-\mathrm{i}T_N\delta t}\mathrm{e}^{-\mathrm{i}H_{el}(\mathbf{R})\frac{\delta t}{2}}, \end{aligned} \tag{6.1}$$

where the second line is an approximation because $[T_N, H_{el}(\mathbf{R})] \neq 0$ in general. The symmetric split of the potential at the beginning and the end of the interval gives an $O(\delta t^3)$ error for small δt. This choice represents an advantage in numerical stability compared to the naive guess $U(\delta t) \approx \mathrm{e}^{-\mathrm{i}H_{el}(\mathbf{R})\delta t}\mathrm{e}^{-\mathrm{i}T_N\delta t}$, which gives an $O(\delta t^2)$ error [3].

Partitioning the propagator in terms of kinetic (T_N) and potential ($H_{el}(\mathbf{R})$) terms has the benefit that T_N acts locally (i.e., is diagonal) in nuclear momentum space whereas $H_{el}(\mathbf{R})$ acts locally in position space. Therefore, the corresponding exponential operators can be implemented as pointwise multiplications on the wavefunction in the respective spaces. As transforming between the position and momentum spaces can be done via the fast Fourier transform (FFT), the SO method can be carried out numerically by applying the potential operator in $\mathbf{R}$ space, Fourier-transforming to $\mathbf{k}$ space, applying the kinetic operator, inverse Fourier-transforming back to the original space, and finally applying the remaining potential operator:

$$U(\delta t) = \mathrm{e}^{-\mathrm{i}H_{el}(\mathbf{R})\frac{\delta t}{2}}\mathrm{FFT}^{-1}\mathrm{e}^{-\mathrm{i}T_N(\mathbf{k})\delta t}\mathrm{FFT}\mathrm{e}^{-\mathrm{i}H_{el}(\mathbf{R})\frac{\delta t}{2}}, \tag{6.2}$$

where $T_N(\mathbf{k}) = \mathbb{I}_{\mathcal{S}} \otimes \sum_i \frac{k_i^2}{2m_i}$, with $\mathbb{I}_{\mathcal{S}}$ being the identity in the system (electronic) subspace, k_i and m_i the conjugate momentum and mass corresponding to nuclear coordinate R_i. A more detailed explanation of how to use this equation will be given in the next sections.

The SO method allows us to propagate a wavepacket under $H_0(\mathbf{R})$. From the previous chapters, we know that once we have access to a series of nuclear wavepackets, we can construct various spectroscopic observables by adding the corresponding wavepacket overlaps in a way that is consistent with phase-matching conditions. In what follows, we give a few examples of how this is done explicitly for several spectroscopies.

6.2 Numerical simulation of frequency-resolved linear absorption

Throughout the book, we have mainly focused on frequency-integrated spectroscopic observables, where the measured signal is not decomposed into frequencies through a spectrometer. However, many of the results derived can be readily generalized to situations where the colors can be separated. These situations are known under the name of *frequency-resolved spectroscopy*, introduced in subsection 3.3 and appendix D, where we also derive the time-domain approach to the calculation of linear absorption spectra $A_{\mathbf{k}_P}(\omega)$. In this section, we study the conceptually simple frequency-resolved linear absorption spectra as the first step towards learning how to simulate more complex spectroscopic signals.

As shown in example 12 in appendix D, $A_{\mathbf{k}_P}(\omega)$ is the spectrum

$$A_{\mathbf{k}_P}(\omega) = \sum_m \frac{4\pi^2\omega}{c} |\langle m|\boldsymbol{\mu}\cdot\mathbf{e}_P|\Psi_0\rangle|^2 \delta(E_m - E_0 - \omega), \tag{6.3}$$

where m sums over all vibronic eigenstates of $H_0(\mathbf{R})$. This expression can be interpreted as the ratio of photon energy at frequency ω that is absorbed by the material with respect to the total incident photon energy at the same frequency. A photon mediates the dipole-allowed transition between the initial and final vibronic eigenstates $|\Psi_0\rangle$ and $|m\rangle$. In complex systems with many nuclear degrees of freedom, there is an approximate continuum of vibrational levels in equation (6.3), which can effectively broaden the delta funtion at a finite spectral resolution. Hence, it is often more convenient and physically intuitive to recast equation (6.3) in the time domain.

For concreteness, let us consider again the simple case of a coupled dimer whose Hamiltonian $H_0(\mathbf{R})$ is given in equations (2.1) and (2.2). Assuming that the initial state of the molecular system before photoexcitation is $|\Psi_0\rangle = |g\rangle|\nu_n^{(g)}\rangle$, that is, the nth vibrational eigenstate of $H_g(\mathbf{R})$, the absorption spectrum can be computed as the Fourier transform of an autocorrelation function (appendix D),

$$\begin{aligned} A_{\mathbf{k}_P}(\omega) &= \frac{2\pi\omega}{c}\int_{-\infty}^{\infty} dt\, e^{i(\omega+E_0)t} \langle\Psi_0|(\boldsymbol{\mu}\cdot\mathbf{e}_P^*)e^{-iH_0t}(\boldsymbol{\mu}\cdot\mathbf{e}_P)|\Psi_0\rangle \\ &= \frac{2\pi\omega}{c}\int_{-\infty}^{\infty} dt\, e^{i(\omega+\omega_{g_n})t}\langle\nu_i^{(g)}|\left(\langle g|e^{iH_0t}(\boldsymbol{\mu}_{ga}\cdot\mathbf{e}_P^*|g\rangle\langle a| + \boldsymbol{\mu}_{gb}\cdot\mathbf{e}_P^*|g\rangle\langle b|\right) \\ &\quad \times e^{-iH_0t}(\boldsymbol{\mu}_{ag}\cdot\mathbf{e}_P|a\rangle\langle g| + \boldsymbol{\mu}_{bg}\cdot\mathbf{e}_P|b\rangle\langle g|)|g\rangle|\nu_n^{(g)}\rangle \\ &= \frac{2\pi\omega}{c}\sum_{i,q=a,b}\int_{-\infty}^{\infty} dt\, e^{i(\omega+\omega_{g_n})t}(\boldsymbol{\mu}_{gi}\cdot\mathbf{e}_P^*)(\boldsymbol{\mu}_{qg}\cdot\mathbf{e}_P)\underbrace{\langle\nu_n^{(g)}|\langle i|e^{-iH_{\mathrm{SEM}}t}|q\rangle|\nu_n^{(g)}\rangle}_{\equiv S_{iq}^n(t)}, \end{aligned} \tag{6.4}$$

where we have used the fact that $E_0 = \omega_{g_n}$ is the eigenenergy of H_0 associated with $|g\rangle|\nu_n^{(g)}\rangle$, expressed the dipole operator in terms of the SEM diabatic states (sites) i, $q{=}a, b$ (equation (2.9)), and replaced the action of H_0 on $|q\rangle$ and $|i\rangle$ with H_{SEM}.[1] Here, we interpret $|q\rangle|\nu_n^{(g)}\rangle$ and $|i\rangle|\nu_n^{(g)}\rangle$ as ground-state nuclear wavepackets vertically excited to the SEM states $|q\rangle$ and $|i\rangle$. Had these wavepackets stayed in $|g\rangle$, they

[1] H_0 is block diagonal; it does not mix states in different electronic excitation manifolds. Therefore, states launched by the light from the GSM to the SEM remain in the latter under unitary evolution via H_0.

would remain stationary, but, being promoted to the SEM, they feel new forces that move them along diabatic surfaces and also due to the nonzero coupling $J(\mathbf{R})$ between them. $S^n_{iq}(t)$ corresponds to the wavepacket overlap between the final propagated state $e^{-iH_{\text{SEM}}t}|q\rangle|\nu_n^{(g)}\rangle$ and the initial state frozen at $|i\rangle|\nu_n^{(g)}\rangle$. The linear absorption spectrum is the Fourier transform of a weighted average of the four possible wavepacket overlaps.

Equation (6.4) shows that we can calculate the linear absorption spectrum without explicit consideration of the optical electric field. We need only its polarization $\mathbf{e}_P$. For nonlinear spectroscopy, we will need to consider the action of the pulses explicitly.

To make our example more concrete, we consider the case of only two nuclear degrees of freedom, $\mathbf{R} = (x, y)$. A numerical implementation of equation (6.4) for this scenario can be found in `dimer_abs.m`, which follows this algorithm:

(1) We define the initial ($t = 0$) two-component wavefunctions representing the state immediately after excitation to either of the electronic states $|a\rangle$ or $|b\rangle$ in the SEM:

$$\psi^{(0)}(x,y|a) = \begin{pmatrix} \langle x,y|\nu_n^{(g)}\rangle \\ 0 \end{pmatrix},$$
$$\psi^{(0)}(x,y|b) = \begin{pmatrix} 0 \\ \langle x,y|\nu_n^{(g)}\rangle \end{pmatrix}. \tag{6.5}$$

Here, $\langle x,y|\nu_n^{(g)}\rangle$ is the coordinate representation of the initial nuclear wavepacket. Each component of the two-component wavefunction represents a vibrational wavepacket located in the diabatic electronic state a or b. For every time t, we define $\psi^{(t)}(x,y|j) = e^{-iH_{\text{SEM}}T}\psi^{(0)}(x,y|j)$ for $j = a, b$. In the MATLAB® code, $H_{\text{GSM}} = H_g(x,y) = T_N + \frac{x^2}{2} + \frac{y^2}{2}$ is a two-dimensional harmonic oscillator (see `PotentialBuilder.m`), so the initial wavepacket acquires the form $\langle x,y|\nu_n^{(g)}\rangle = \psi_{n_1}(x)\psi_{n_2}(y)$, a product of two independent harmonic oscillator states with quantum numbers n_1 and n_2. This wavefunction can be represented as $N_x \times N_y$ matrices by sampling points, so each two-components wavefunction involves either two of these matrices or, as in the MATLAB® code, a single $2N_x \times N_y$ matrix. If the GSM eigenstates are not known analytically, they can be computed by solving the time-independent Schrödinger equation numerically.

(2) Using the split-operator method, we compute $\psi^{(t)}(x,y|j)$ for all $t > 0$ by repeatedly applying $U(\delta t) = e^{-iH_{\text{SEM}}\delta t}$ to $\psi^{(0)}(x,y|j)$. Restricting our attention to the SEM and using the two-component wavefunction notation, the different terms in H_{SEM} read,

$$H_{el}(x,y) = \begin{pmatrix} V_a(x,y) & J(x,y) \\ J(x,y) & V_b(x,y) \end{pmatrix}, \tag{6.6a}$$

$$T_N(k_x, k_y) = \begin{pmatrix} \frac{1}{2}\left(k_x^2 + k_y^2\right) & 0 \\ 0 & \frac{1}{2}\left(k_x^2 + k_y^2\right) \end{pmatrix}, \tag{6.6b}$$

where the mass of the nuclei has been folded into the definitions of the momenta k_x and k_y. The corresponding exponential operators for equation (6.2) are,

$$e^{-iT_N(k_x, k_y)\delta t} = \begin{pmatrix} e^{-i\frac{1}{2}\left(k_x^2+k_y^2\right)\delta t} & 0 \\ 0 & e^{-i\frac{1}{2}\left(k_x^2+k_y^2\right)\delta t} \end{pmatrix}, \tag{6.7a}$$

$$e^{-iH_{el}(x,y)\frac{\delta t}{2}} = \exp\left[-iW(x,y)\frac{\delta t}{4}\right]\left[\cos\left[\sqrt{D(x,y)}\frac{\delta t}{4}\right]\begin{pmatrix} 1 & 0 \\ 0 & 1 \end{pmatrix} + i\frac{\sin\left[\sqrt{D(x,y)}\frac{\delta t}{4}\right]}{\sqrt{D(x,y)}}\begin{pmatrix} \Delta(x,y) & -2J(x,y) \\ -2J(x,y) & -\Delta(x,y) \end{pmatrix}\right], \tag{6.7b}$$

where we have analytically evaluated the matrix exponential $e^{-iH_{el}(x,y)\frac{\delta t}{2}}$, with the parameters in the expression being

$$W(x,y) = V_a(x,y) + V_b(x,y), \tag{6.8}$$

$$\Delta(x,y) = V_b(x,y) - V_a(x,y), \tag{6.9}$$

$$D(x,y) = 4J(x,y)^2 + \Delta(x,y)^2. \tag{6.10}$$

A more detailed explanation of this propagation step will be given in the paragraphs following this outline of the algorithm.

(3) Numerically calculate the four different overlaps $S_{ip}^n(t)$ at every $t \geqslant 0$ using matrix pointwise multiplications.
(4) Extend the time-dependent signal to negative times using $S_{pi}^n(-t) = S_{ip}^n(t)^*$.
(5) Evaluate $S_{ip}^n(t)$ in equation (6.4) using a fast Fourier transform (FFT) routine. If the time grid consists of $T/\delta t$ points spaced by δt, the frequency grid takes on values in $[-\pi/\delta t, \pi/\delta t)$ spaced by $\delta\omega = 2\pi/T$. This is intuitively clear: a fine grid in time (small δt) allows us to capture higher-frequency contributions, up to $\pi/\delta t$. Likewise, a longer total simulation time T increases the frequency resolution $\delta\omega$.

From a practical point of view, experimental limitations and various types of line broadening mean that one is often only interested in the coarse structure of the absorption spectrum, i.e., that only a short time-evolution is necessary. As Heller has pointed out [6], a short simulation already informs us on the early dynamical processes that contribute to the coarse structure of an absorption spectrum. This often provides more physical insight than an exhaustive enumeration of thousands of eigenvalues and eigenstates that may be required to adequately describe a complex molecular system with many nuclear degrees of freedom.

(6) To model finite temperature, average $A_{\mathbf{k}_P}(\omega)$ over an ensemble of initial vibronic eigenstates of the GSM $|\Psi_0\rangle = |g\rangle|\nu_n^{(g)}\rangle$ chosen according to a thermal distribution $p_n = \exp(-\omega_{g_n}/k_B\mathscr{T})/\mathrm{Tr}[\exp(-H_g/k_B\mathscr{T})]$. In the case of many eigenstates within $k_B(\mathscr{T})$, this average can be carried out by Monte Carlo sampling.

(7) A second layer of averaging is required if the simulation corresponds to isotropically distributed molecules in solution. The resulting spectrum can be calculated by averaging three different signals in which the molecular coordinates are fixed and the pulse polarization $\mathbf{e}_P$ is varied along the **x**, **y** and **z** axes, see appendix F. The different layers of this averaging can also be optimized by Monte Carlo strategies.

Now that we have a strategy to numerically implement equation (6.4), we address further technical details on how to carry out the specific operation $\psi^{(t+\delta t)}(x, y|j) = U(\delta t)\psi^{(t)}(x, y|j)$. The starting point is the general wavefunction $\psi^{(t)}(x, y|j) = \begin{pmatrix} \psi_a^{(t)}(x, y|j) \\ \psi_b^{(t)}(x, y|j) \end{pmatrix}$ at time t. The explicit steps for propagating this wavefunction to $t + \delta t$ are the following (see the MATLAB® functions `one_time_step_*.m`):

1. Apply the potential operator $e^{-iH_{el}(x,y)\frac{\delta t}{2}}$:

$$
\begin{aligned}
\psi_{\mathrm{temp}}^{(t)}(x, y|j) &= e^{-iH_{el}(x,y)\frac{\delta t}{2}}\psi^{(t)}(x, y|j) \\
&= \exp\left[-iW(x, y)\frac{\delta t}{4}\right] \\
&\quad \times \left[\begin{pmatrix} \cos\left[\sqrt{D(x, y)}\frac{\delta t}{4}\right]\psi_a^{(t)}(x, y|j) \\ \cos\left[\sqrt{D(x, y)}\frac{\delta t}{4}\right]\psi_b^{(t)}(x, y|j) \end{pmatrix} + i\frac{\sin\left[\sqrt{D(x, y)}\frac{\delta t}{4}\right]}{\sqrt{D(x, y)}}\right. \\
&\quad \left. \times \begin{pmatrix} \Delta(x, y)\psi_a^{(t)}(x, y|j) - 2J(x, y)\psi_b^{(t)}(x, y|j) \\ -2J(x, y)\psi_a^{(t)}(x, y|j) - \Delta(x, y)\psi_b^{(t)}(x, y|j) \end{pmatrix}\right],
\end{aligned}
$$

which, we notice, involves pointwise multiplications in the real space grid and exchanging of the different components of the spinor. Obviously, if J vanishes, this step does not involve exchange of the components.

2. Fourier transform the resulting wavefunction,

$$\psi^{(t)}_{\text{temp}}(k_x, k_y|j) = \sum_{x,y} \psi^{(t)}_{\text{temp}}(x, y|j)\mathrm{e}^{-\mathrm{i}(xk_x+yk_y)} = \begin{pmatrix} \sum_{x,y} \psi^{(t)}_{\text{temp},a}(x, y|j)\mathrm{e}^{-\mathrm{i}(xk_x+yk_y)} \\ \sum_{x,y} \psi^{(t)}_{\text{temp},b}(x, y|j)\mathrm{e}^{-\mathrm{i}(xk_x+yk_y)} \end{pmatrix}. \quad (6.11)$$

Here, if the grid spacing for x is Δ_x, k_x takes on N_x values $k_x \in [-\frac{\pi}{\Delta_x}, \frac{\pi}{\Delta_x})$ with spacing $\Delta_{k_x} = \frac{2\pi}{N_x \Delta_x}$. Similar conditions follow for y and k_y. The Fourier transform is achieved using an FFT routine.[2]

3. Apply the kinetic energy propagator, which is purely multiplicative in momentum space,

$$\varphi^{(t)}_{\text{temp}}(k_x, k_y|j) = \mathrm{e}^{-\mathrm{i}T_N(k_x,k_y)\delta t}\psi^{(t)}_{\text{temp}}(k_x, k_y|j) = \begin{pmatrix} \mathrm{e}^{-\mathrm{i}\left(\frac{k_x^2+k_y^2}{2}\right)\delta t}\psi^{(t)}_{\text{temp},a}(k_x, k_y|j) \\ \mathrm{e}^{-\mathrm{i}\left(\frac{k_x^2+k_y^2}{2}\right)\delta t}\psi^{(t)}_{\text{temp},b}(k_x, k_y|j) \end{pmatrix}.$$

4. Fourier transform back to coordinate space $\varphi^{(t)}_{\text{temp}}(x, y|j) = \frac{1}{N_x N_y}\sum_{x,y}\varphi^{(t)}_{\text{temp}}(k_x, k_y|j)\mathrm{e}^{\mathrm{i}(xk_x+yk_y)}$ using the inverse FFT.
5. Obtain the final result for one propagation step by applying the potential propagator again, $\psi^{(t+\delta t)}(x, y|j) = \mathrm{e}^{-\mathrm{i}H_{el}(x,y)\frac{\delta t}{2}}\varphi^{(t)}_{\text{temp}}(x, y|j)$.

Repeating for many δt allows simulation of the linear absorption spectrum from equation (6.4).

6.3 Numerical simulation of frequency-integrated linear and nonlinear spectra

The main emphasis of this book has been on frequency-integrated spectroscopies. The previous section described frequency-resolved linear spectra and provided us with the wavepacket methodology that will be exploited in the present section.

[2] Using the FFT phase convention in MATLAB®, the built-in function `fftshift` is helpful to rearrange the FFT output for $k \in [-\pi/\Delta_x, \pi/\Delta_x)$. For other numerical packages, care must be taken in ordering the output of the FFT appropriately.

We are now concerned with the simulation of arbitrary frequency-integrated signals S_s, which we rewrite from equation (3.52),

$$S_s = 2\Im \int_{-\infty}^{\infty} dt' \varepsilon_{\mathrm{LO}}^*(t' - t_{\mathrm{LO}}) e^{-i\phi_{\mathrm{LO}}} \mathbf{e}_{\mathrm{LO}}^* \cdot \mathbf{P}_{\mathbf{k}_s}(t') e^{i\phi_s}. \tag{6.12}$$

The techniques we now present for frequency-integrated spectra can be readily adapted for the computation of frequency-resolved spectra (see appendix D), as mentioned at the end of section 3.3.

The evaluation of equation (6.12) is straightforward once the perturbative polarization $\mathbf{P}_{\mathbf{k}_s}(t')$ is obtained, which, in turn, is a matter of wavefunction propagation. In the previous section, the propagation $U(\delta t)$ involved the SEM only. For nonlinear spectroscopy, propagation of wavepackets in the GSM and the DEM is also needed. However, we have already covered the most difficult part, at least within the model Hamiltonian $H_0(\mathbf{R})$ (equation (2.1)). We limit ourselves to the case where both $H_0(\mathbf{R})$ and $U(\delta_T)$ are block diagonal, not coupling wavepackets in different excitation manifolds in the absence of optical perturbations. Furthermore, in our coupled-dimer model, the GSM and the DEM consist of a single electronic state each. Hence, the two-component notation above is not needed for those manifolds, and for them, each step of the split-operator method can be implemented as a pointwise multiplication.

In order to compute $\mathbf{P}_{\mathbf{k}_s}(t')$, we must implement the optically induced transitions between different manifolds in a numerically stable and tractable way. That is, we need to calculate the various perturbative wavefunctions using the total Hamiltonian $H = H_0 + V(\mathbf{r}, t)$, where the optical perturbations are contained in $V(\mathbf{r}, t)$ (equations (3.1)–(3.4)). The propagation due to $H_0(R)$ is nonperturbative, requiring only a sufficiently short δt for numerical accuracy. The perturbation is of the form $V(\mathbf{r}, t) = -\sum_n \boldsymbol{\mu} \cdot \mathbf{e}_n \varepsilon_n(t - t_n) e^{i\mathbf{k}_n \cdot \mathbf{r} + i\phi_n}$, coupling the molecular dipole with the electric field of the light pulses.

We propagate multiple wavefunctions according to H_0 and include the perturbations at various orders in H' (or equivalently, in the field strength η, see equation (3.4). To do so, we discretize the integrals corresponding to various wavefunction perturbations. Each interaction with the optical field takes a wavefunction to a higher-order wavefunction; this can take place in two ways, for the different rotations in the RWA:

$$\begin{aligned} |\Psi_{\ldots \pm i+n}(t)\rangle &= i \int_{-\infty}^{t} dt' e^{-iH_0(t-t')} \boldsymbol{\mu} \cdot \mathbf{e}_n \varepsilon_n(t' - t_n) |\Psi_{\ldots \pm i}(t')\rangle \\ &\approx i\delta t \sum_{t' \leqslant t} U(t - t') \boldsymbol{\mu} \cdot \mathbf{e}_n \varepsilon_n(t' - t_n) |\Psi_{\ldots \pm i}(t')\rangle, \end{aligned} \tag{6.13}$$

$$\begin{aligned} |\Psi_{\ldots \pm i-n}(t)\rangle &= i \int_{-\infty}^{t} dt' e^{-iH_0(t-t')} \boldsymbol{\mu} \cdot \mathbf{e}_n^* \varepsilon_n^*(t' - t_n) |\Psi_{\ldots \pm i}(t')\rangle \\ &\approx i\delta t \sum_{t' \leqslant t} U(t - t') \boldsymbol{\mu} \cdot \mathbf{e}_n^* \varepsilon_n^*(t' - t_n) |\Psi_{\ldots \pm i}(t')\rangle, \end{aligned} \tag{6.14}$$

where the sum is over discrete times t' separated by δt, $U(t-t') = e^{-iH_0(t-t')}$ is the propagator of the molecular Hamiltonian, and we have followed the wavefunction subscript convention introduced in section 3.2 and used throughout the book. Equations (6.13) and (6.14) suggest a numerical implementation where one stores as many wavefunctions as times t' in the simulation and subsequently adds the results up to time t. A simpler implementation exploits the following recursion relations (see example 11),

$$|\Psi_{\dots\pm i+n}(t)\rangle = U(\delta t)|\Psi_{\dots\pm i+n}(t-\delta t)\rangle + i\delta t\boldsymbol{\mu}\cdot\mathbf{e}_n\varepsilon_n(t-t_n)|\Psi_{\dots\pm i}(t)\rangle, \tag{6.15}$$

$$|\Psi_{\dots\pm i-n}(t)\rangle = U(\delta t)|\Psi_{\dots\pm i-n}(t-\delta t)\rangle + i\delta t\boldsymbol{\mu}\cdot\mathbf{e}_n^*\varepsilon_n^*(t-t_n)|\Psi_{\dots\pm i}(t)\rangle. \tag{6.16}$$

These relations give a simple method for evolving $|\Psi_{\dots\pm i\pm n}(t-\delta t)\rangle$ to $|\Psi_{\dots\pm i\pm n}(t)\rangle$ for a time step δt. First, a SO propagation via $U(\delta t)$ is carried out. The result is then supplemented with fresh perturbative amplitude from the 'source' wavefunction $|\Psi_{\dots\pm i}(t)\rangle$, weighted by the corresponding pulse amplitude and transition dipole moment. The algorithm can be simplified if the pulse amplitudes are nonnegligible only during a certain time window, before and after which the propagation can be regarded as due to $U(\delta t)$ alone.

The only missing ingredient is an initial condition, so we set $|\Psi_0(t_0)\rangle = |g\rangle|\nu_n^{(g)}\rangle$ and $|\Psi_{\dots\pm i}(t_0)\rangle = 0$ otherwise. This choice means that at the beginning of the simulation, only the zeroth wavefunction is populated in a vibrational eigenstate of H_{GSM}, and the action of pulses populates the perturbative wavefunctions accordingly, whereas $|\Psi_0(t)\rangle$ simply evolves under H_{GSM}. This technique works for linear spectroscopy as presented in section 4.1, but its main value lies in predicting nonlinear signals. We originally devised it to carry out our computational simulations in [15].

Note that, due to phase-matching, only a subset of perturbative wavefunctions needs to be computed to obtain the polarization $\mathbf{P}_{\mathbf{k}_s}$. Also, the block off-diagonal form of the dipole operator and the RWA imply that these wavefunctions reside in *only one* of the excitation manifolds. Hence, for each required perturbative wavefunction, we only need to allocate memory for one (or possibly a few) nuclear wavepackets that do not vanish. This will become clearer in the next subsection.

Example 11. Proof of algorithm for wavefunction propagation under optical perturbations

Prove that equations (6.15) and (6.13) are indeed equivalent. The equivalence of equations (6.16) and (6.14) follow analogously.

Solution

We proceed with a proof by induction on t. First, suppose that $|\Psi_{\dots\pm i+n}(t)\rangle$ is correctly given by equation (6.13). We want to show that the recurrence relation (6.13) will yield the correct form of $|\Psi_{\dots\pm i+n}(t+\delta t)\rangle$,

$$U(\delta t)|\Psi_{\cdots\pm i+n}(t)\rangle + i\delta t\boldsymbol{\mu}\cdot\mathbf{e}_n\varepsilon_n(t+\delta t-t_n)|\Psi_{\cdots\pm i}(t+\delta t)\rangle$$
$$= i\delta t\sum_{t'\leqslant t}U(\delta t)U(t-t')\boldsymbol{\mu}\cdot\mathbf{e}_n\varepsilon_n(t'-t_n)|\Psi_{\cdots\pm i}(t')\rangle$$
$$+ i\delta t\boldsymbol{\mu}\cdot\mathbf{e}_n\varepsilon_n(t+\delta t-t_n)|\Psi_{\cdots\pm i}(t+\delta t)\rangle$$
$$= i\delta t\sum_{t'\leqslant t+\delta t}U(t+\delta t-t')\boldsymbol{\mu}\cdot\mathbf{e}_n\varepsilon_n(t'-t_n)|\Psi_{\cdots\pm i}(t')\rangle$$
$$= |\Psi_{\cdots\pm i+n}(t+\delta t)\rangle,$$

where we have used the composition property $U(\delta t)U(t-t') = U(t+\delta t-t')$ and the fact that $U(0) = 1$. The only missing ingredient is the initial condition of the wavefunctions, which have been correctly set by the definitions above.

Numerical calculation of nonlinear spectroscopic signals: transient gratings, pump probe, photon echo and two-dimensional electronic spectra

We consider again the coupled dimer described by the molecular Hamiltonian $H_0(\mathbf{R})$ and investigate how to numerically compute various nonlinear spectroscopic signals of such a molecule starting from a detailed analysis of its TG spectrum (example 10). Once we understand this example, the other spectroscopic signals will be straightforwardly generated. The relevant expressions for TG spectroscopy are equations (5.28) and (5.34). Rewriting them here, $S_{\mathrm{TG}}[\gamma]$ is the TG signal of interest, and it can be computed as an overlap between a third-order polarization along the $\mathbf{k}_{\mathrm{TG}}$ direction and the LO pulse 4,

$$S_{\mathrm{TG}}[\gamma] = 2\Im\int_{-\infty}^{\infty}\mathrm{d}t'\varepsilon_4^*(t'-t_{P'})\mathbf{e}_4^*\cdot\mathbf{P}^{(3)}_{\mathbf{k}_{TG}}(t')\mathrm{e}^{\mathrm{i}\gamma}. \tag{6.17}$$

Here, the TG polarization is given by

$$P^{(3)}_{\mathbf{k}_{\mathrm{TG}}}(t) = \langle\Psi_{+1}(t')|\boldsymbol{\mu}|\Psi_{+2+3}(t')\rangle + \langle\Psi_{+1-3}(t')|\boldsymbol{\mu}|\Psi_{+2}(t')\rangle$$
$$+ \langle\Psi_{+1-2}(t')|\boldsymbol{\mu}|\Psi_{+3}(t)\rangle + \langle\Psi_0(t)|\boldsymbol{\mu}|\Psi_{+2-1+3}(t')\rangle \tag{6.18}$$

and $\gamma = -\phi_{12} + \phi_{34}$ is a combination of phases from the various pulses (equation (5.29)). The strategy is to compute the various wavefunctions involved in $P^{(3)}_{\mathbf{k}_{TG}}(t)$. Recall that $t_1 = t_2 = t_P$ and $t_3 = t_4 = t_{P'}$, and $T = t_{P'} - t_P \gg \sigma_i$ for every pulse i. It is this time ordering that allows us to neglect contributions such as $\langle\Psi_0(t)|\boldsymbol{\mu}|\Psi_{+3+2-1}(t')\rangle$, where pulse 3 acts before the other two pulses.

In the RWA, $|\Psi_0(t)\rangle$ is a GSM wavefunction; $|\Psi_{+1}(t)\rangle$, $|\Psi_{+2}(t)\rangle$, $|\Psi_{+3}(t)\rangle$ are SEM states; $|\Psi_{+2+3}(t)\rangle$ is a DEM state; $|\Psi_{+1-2}(t)\rangle$ and $|\Psi_{+1-3}(t)\rangle$ are in the GSM; and $|\Psi_{+2-1+3}(t)\rangle$ is in the SEM. Even though the GSM state $|\Psi_{+2-1}(t)\rangle$ does not feature in equation (6.18), it is needed to generate $|\Psi_{+2-1+3}(t)\rangle$ via pulse 3, so we also need to compute it. This gives nine wavefunctions to compute. Specializing again to two nuclear degrees of freedom x and y, they are:

- Four GSM wavepackets $\psi^{(t)}_{n,\,\mathrm{GSM}}(x, y)$ for $n = 0$, $+1-2$, $+1-3$, $+2-1$. $\psi^{(t)}_{0,\,\mathrm{GSM}}$, being an eigenstate of $H_0(\mathbf{R})$, has a trivial evolution, which, if desired, can be analytically calculated instead of being propagated by the split-operator method.
- Four SEM two-component wavefunctions $\psi^{(t)}_{n,\,\mathrm{SEM}}(x, y) = \begin{pmatrix} \psi_{+n,\mathrm{SEM},a}(x, y) \\ \psi_{+n,\mathrm{SEM},b}(x, y) \end{pmatrix}$ for $n = 1, 2, 3, +2-1+3$.
- One DEM wavepacket $\psi^{(t)}_{+2+3,\,\mathrm{DEM}}(x, y)$.

This method is used in `dimer_tg.m`. For a fixed waiting time T, the propagation of these wavefunctions and the calculation of the subsequent TG signal proceeds as follows,

- Let $t = 0$ to be the center time of the first pulse t_1. Begin the simulation at the negative time $t_0 = -\kappa \max(\sigma_1, \sigma_2)$ when all perturbative fields are weak; we choose $\kappa \approx 3$.[3] This choice of t_0 allows us to simulate most of the envelope of pump pulses 1 and 2. For the TG setup, we choose $t_1 = t_2 = 0$ and $t_3 = t_4 = T$. The initial condition is $\psi^{(t_0)}_{0,\,\mathrm{GSM}} = \langle x, y | \nu^{(g)}_n \rangle$, the initial vibronic eigenstate of the GSM, whereas the other eight wavefunctions begin as zero matrices.
- Evolve the nine wavepackets. For example,
  - Evolve $\psi^{(t)}_{0,\,\mathrm{GSM}}(x, y)$ under H_0, or equivalently, under H_{GSM},

$$\psi^{(t)}_{0,\,\mathrm{GSM}}(x, y) = \mathrm{e}^{-\mathrm{i}E_0 t}\psi^{(0)}_{0,\,\mathrm{GSM}}(x, y).$$

  - The other wavefunctions are evolved by the two-step approach of simulating excitation and propagating under the molecular Hamiltonian using the split-operator method. As an illustration,
    * One time step for the evolution of $\psi^{(t)\mathrm{temp}}_{+2-1,\,\mathrm{GSM}}(x, y)$ consists of (a) split-operator propagation under H_0 (H_{GSM} in this case), and (b) amplitude addition from $\psi^{(t)}_{+2,\,\mathrm{SEM}}(x, y)$,

$$\psi^{(t+\delta t)\mathrm{temp}}_{+2-1,\,\mathrm{GSM}}(x, y) = \mathrm{e}^{-\mathrm{i}H_{\mathrm{GSM}}\delta t}\psi^{(t)\mathrm{temp}}_{+2-1,\,\mathrm{GSM}}(x, y)\ [\text{using SO method}]. \tag{6.19a}$$

$$\begin{aligned}\psi^{(t+\delta t)}_{+2-1,\,\mathrm{GSM}}(x, y) &= \psi^{(t+\delta t)\mathrm{temp}}_{+2-1,\,\mathrm{GSM}}(x, y) + \mathrm{i}\,\delta t\,\boldsymbol{\mu}\cdot\mathbf{e}_1^*\varepsilon_1^*(t+\delta t - t_1)\psi^{(t+\delta t)}_{+2,\,\mathrm{SEM}}(x, y) \\ &= \psi^{(t+\delta t)\mathrm{temp}}_{+2-1,\,\mathrm{GSM}}(x, y) + \mathrm{i}\,\delta t\,\varepsilon_1^*(t+\delta t - t_1)\Big\{\boldsymbol{\mu}_{ga}\cdot\mathbf{e}_1^*\psi^{(t+\delta t)}_{+2,\,\mathrm{SEM},\,a}(x, y) \\ &\quad + \boldsymbol{\mu}_{gb}\cdot\mathbf{e}_1^*\psi^{(t+\delta t)}_{+2,\,\mathrm{SEM},\,b}(x, y)\Big\},\end{aligned} \tag{6.19b}$$

[3] Note the different convention of initial time with respect to the assumption in equation (3.9).

* Similarly, one time step for the evolution of $\psi^{(t)\text{temp}}_{+2-1+3,\,\text{SEM}}(x, y)$ is given by

$$\psi^{(t+\delta t)\text{temp}}_{+2-1+3,\,\text{SEM}}(x, y) = \mathrm{e}^{-\mathrm{i}H_{\text{SEM}}\delta t}\psi^{(t)}_{+2-1+3,\,\text{SEM}}(x, y) \text{ [using SO method]}. \tag{6.20a}$$

$$\begin{aligned}
\psi^{(t+\delta t)}_{+2-1+3,\,\text{SEM}}(x, y) &= \psi^{(t+\delta t)\text{temp}}_{+2-1+3,\,\text{SEM}}(x, y) + \mathrm{i}\,\delta t\boldsymbol{\mu}\cdot\mathbf{e}_3\varepsilon_3(t+\delta t-t_3)\psi^{(t+\delta t)}_{+2-1,\,\text{GSM}}(x, y) \\
&= \begin{pmatrix} \psi^{(t)\text{temp}}_{+2-1+3,\,\text{SEM},\,a}(x, y) \\ \psi^{(t)\text{temp}}_{+2-1+3,\,\text{SEM},\,b}(x, y) \end{pmatrix} \\
&\quad + \begin{pmatrix} \mathrm{i}\,\delta t\boldsymbol{\mu}_{ag}\cdot\mathbf{e}_3\varepsilon_3(t+\delta t-t_3)\psi^{(t+\delta t)}_{+2-1,\,\text{GSM}}(x, y) \\ \mathrm{i}\,\delta t\boldsymbol{\mu}_{bg}\cdot\mathbf{e}_3\varepsilon_3(t+\delta t-t_3)\psi^{(t+\delta t)}_{+2-1,\,\text{GSM}}(x, y) \end{pmatrix},
\end{aligned} \tag{6.20b}$$

* As a final example, we show one time step for the evolution of $\psi^{(t)\text{temp}}_{+2+3,\,\text{DEM}}(x, y)$:

$$\psi^{(t+\delta t)\text{temp}}_{+2+3,\,\text{DEM}}(x, y) = \mathrm{e}^{-\mathrm{i}H_{\text{DEM}}\delta t}\psi^{\text{temp}(t)}_{+2+3,\,\text{DEM}}(x, y) \text{ [using SO method]}. \tag{6.21a}$$

$$\begin{aligned}
\psi^{(t+\delta t)}_{+2+3,\,\text{DEM}}(x, y) &= \psi^{(t+\delta t)\text{temp}}_{+2+3,\,\text{DEM}}(x, y) \\
&\quad + \mathrm{i}\,\delta t\boldsymbol{\mu}\cdot\mathbf{e}_3\varepsilon_3(t+\delta t-t_3)\psi^{(t+\delta t)}_{+2,\,\text{SEM}}(x, y) \\
&= \psi^{(t+\delta t)\text{temp}}_{+2+3,\,\text{DEM}}(x, y) \\
&\quad + \mathrm{i}\,\delta t\,\varepsilon_3(t+\delta t-t_3)\{\boldsymbol{\mu}_{fa}\cdot\mathbf{e}_3\psi^{(t+\delta t)}_{+2,\,\text{SEM},\,a}(x, y) \\
&\quad + \boldsymbol{\mu}_{fb}\cdot\mathbf{e}_3\psi^{(t+\delta t)}_{+2,\,\text{SEM},\,b}(x, y)\},
\end{aligned} \tag{6.21b}$$

- Evolve wavefunctions until the final time $t_f = T + \kappa\sigma_3$ after the final pulse is effectively done. Just as with the starting time t_0, the additional time after T allows us to simulate most of the envelope of pulse 3 for $t > T$.
- Numerically compute the wavepacket overlaps needed for $\mathbf{P}^{(3)}_{\mathbf{k}_{TG}}\cdot\mathbf{e}^*_4(t)$ (equation (6.18)) using pointwise matrix multiplications.
- Evaluate equation (6.17) for a given choice of γ by numerically computing the overlap between $\mathbf{P}^{(3)}_{\mathbf{k}_{TG}}\cdot\mathbf{e}^*_4(t)$ and $\varepsilon^*_4(t - t_4)$.
- Repeat the entire procedure for a sufficiently dense set of T points in order to resolve the waiting time dynamics of interest (see previous section for discussions of frequency resolution due to finite sampling).
- As in the case of linear absorption, include finite temperature effects by Monte Carlo sampling a thermal ensemble of initial states $|\Psi_0(t_0)\rangle$. If the sample is isotropically distributed in solution, carry out an isotropic average of the signal (see appendix F).

There are many places where one can optimize this algorithm. When the pulses are short, the propagation substeps which involve addition of amplitude via some optical pulse can be omitted unless the time t is near the pulse center. The same is true for $\mathbf{P}^{(3)}_{\mathbf{k}_{TG}} \cdot \mathbf{e}^*_4(t)$, which may be computed only for times close to t_4, and similarly for the overlap integral in equation (6.17), whose integrand is significant only for that same time interval. Finally, it can be expensive to simulate $S_{PP'}(T)$ for various waiting times T by always starting from t_0. One can partially reuse the evolution of the wavefunctions involved in $S_{PP'}(T)$ for the calculation of a successive waiting time, $S_{PP'}(T + \delta T)$, by saving the wavefunctions, say, up to $T - \kappa\sigma_3$, before a considerable amount of pulse 3 is on. The code provided with the book does not exploit these techniques, in order to maintain conceptual simplicity. The presented algorithm was developed originally to tackle the calculations presented in the study of [15], which discusses the witness to distinguish between vibrational and electronic coherences.

The previous method can be readily adapted for the calculation of related spectra:

1. As shown in example 10, when pulses 1 and 2 are identical, and 3 and 4 are too, $S_{\mathrm{TG}}[0](T) = S_{PP'}(T)$, i.e., the TG spectrum reduces to the PP' spectrum. This means that a code to compute TG spectra also computes PP' spectra.
2. A photon-echo (PE) signal is also a straightforward generalization of the TG signal where $t_1 \neq t_2$ and $t_3 \neq t_4$, in general. One can then introduce the *coherence* time $\tau \equiv t_2 - t_1$, *waiting* time $T = t_3 - t_2$, and *echo* time $\bar{t} = t_4 - t_3$ (appendix E). Note that if $\tau \gg \sigma_1, \sigma_2$, the contribution $\langle \Psi_0(t) | \boldsymbol{\mu} | \Psi_{+2-1+3}(t') \rangle$ can be ignored, as it corresponds to pulse 2 acting before pulse 1. Hence, neither $|\Psi_{+2-1}(t')\rangle$ nor $|\Psi_{+2-1+3}(t')\rangle$ need to be computed since they are negligible, and the number of wavefunctions to be propagated is reduced from nine to seven. Finally, since now we are also interested in the polarization signal for $\bar{t} > 0$, we choose the final simulation time t_f to be $t_4 + \kappa\sigma_4$ rather than $t_3 + \kappa\sigma_3$. The rest of the calculation proceeds identically to the TG one, except that the PE signal $S_{\mathrm{PE}}(\tau, T, \bar{t})$ depends on the sampling of three different time intervals.
3. A slight variation of the PE signal is the two-dimensional electronic spectrum (2D-ES), which is the two-dimensional Fourier transform of the complex-valued signal $\Sigma_{\mathrm{PE}}(\tau, T, \bar{t}) = \frac{1}{2}\left\{S_{\mathrm{PE}}[0](\tau, T, \bar{t}) + \mathrm{i}S_{\mathrm{PE}}\left[-\frac{\pi}{2}\right](\tau, T, \bar{t})\right\}$ along the τ and $\bar{t}$ time intervals. This signal is equivalent to $\Sigma_{\mathrm{PE}}(\tau, T, \bar{t}) = -\mathrm{i}\int_{-\infty}^{\infty} \mathrm{d}t' \varepsilon^*_4(t' - t_{P'})\mathbf{e}^*_4 \cdot \mathbf{P}^{(3)}_{\mathbf{k}_{\mathrm{PE}}}(t')$, without the phase γ (appendix E, equations (E.9*a*) and (E.9*b*)). Therefore, the 2D-ES is readily available once the nonlinear polarization $\mathbf{P}^{(3)}_{\mathbf{k}_{\mathrm{PE}}}(t')$ is computed. See the example MATLAB® code `dimer_2DES.m`.
4. The described computational method does not take advantage of the simpler wavepacket overlap expressions for the PP' signal as in equation (4.35). We invite the reader to modify the proposed algorithm to make use of such formulas. Recall that even though they are more compact than a systematic calculation of the nonlinear polarization, they are limited to PP'. The algorithm presented in this section can serve to compute TG, PP', PE or 2D-ES signals at the same time.

5. The method can be readily adapted to simulate frequency-resolved spectra (see appendix D and, specifically, example 14 for details). It can certainly be used for linear spectroscopy, but, as shown in example 12, the explicit inclusion of pulses is not needed in that specific case, and the algorithm presented in 6.2 suffices and is more economical to implement than the present one. The explicit inclusion of pulses is, however, necessary for nonlinear spectra.

The described technique is, by construction, exact within time-dependent perturbation theory. Therefore, before using this method (and most of the calculations offered in this book), one must first confirm that the strength of the pulses lends itself to a perturbative treatment (appendix A). We note that if every pulse $\varepsilon_n(t - t_n)$ is scaled by a constant c, the sth order wavefunction will be scaled by c^s, but its time-dependence will be otherwise left intact. We can use this property to our advantage by artificially increasing the magnitude of the perturbation to be on the same scale as H_0 in order to avoid underflow issues in double-precision arithmetic, which can be caused by computing small perturbative effects in the presence of a large ground-state background. If we are interested in nonperturbative effects of strong fields, the present method will fail, and we will need to resort to other numerical techniques reported in the literature to evolve Hamiltonians that explicitly include the various pulses [4, 8, 13].

6.4 Extensions: boundary conditions and relaxation dynamics

A few extensions to the technique bear mentioning. The first is the effect of a propagated wavepacket reaching the edge of the grid in position space. This has several undesirable effects depending on the exact propagator used, including reflection or, in the case of the split operator, 'wrapping around' and reappearing on the other side as a consequence of the periodicity of the FFT routine. This can introduce major numerical errors, and thus it is essential that the wavepackets remain away from the edges of the simulation region. For some PES, such as the harmonic one, this is relatively simple as the eigenstates are bound and the only requirement is that the simulation region be made large enough. For potentials that have dissociative features, it is either not possible to localize the states or the required simulation region becomes prohibitively large. However, as the spectroscopic signals of interest generally correspond to overlap integrals involving at least one bound state, these wavepackets in the continuum usually contribute very little to the signal of interest. In these cases, we can use an *absorbing boundary* to limit the wavepacket dynamics to the localized portion. If we recall the form of the propagation operator $U(\delta t) = \mathrm{e}^{-\mathrm{i}H_0(\mathbf{R})\delta t}$, it is easy to see that a negative imaginary potential added to the Hamiltonian will result in an exponential decay of the wavefunction in time [9]. Thus, adding a negative imaginary potential around the edges of the simulation region will 'absorb' the wavefunction before it interacts with the boundary, preventing reflections or wraparound. Though a full discussion of absorbing boundaries is beyond the scope of this book, we make two pertinent observations. First, like with a real valued barrier, a wavefunction can reflect off an absorbing potential. This reflection scales with the steepness of the absorbing barrier, so a gentle increase of such a barrier over a longer region will reduce errors resulting from partial reflection [1]. Second, it has been shown that it is in general superior to use a *complex absorbing potential* [12]. This is because low frequency waves are poorly absorbed by a purely imaginary potential, in analogy to electromagnetic waves inside a metal. Adding a real potential with negative slope accelerates the wavepacket as it falls and absorbs it more efficiently.

A second extension considers dealing implicitly with the vibrational bath $\mathscr{B}$ via the theory of open quantum systems (OQS). This might be relevant when dealing with complex molecular systems with many vibrational degrees of freedom, where we wish to explicitly simulate none or only a few of the nuclear degrees of freedom, while the rest act as a thermal bath. In such a case, the electronic degrees of freedom follow nonunitary evolution featuring dephasing and relaxation processes (see chapter 1). A possible way to extend the numerical methodology for wavepackets to this problem is the Monte-Carlo wavefunction formalism [2]. This technique is used for including relaxation and dephasing dynamics while simultaneously maintaining the low-dimensional structure of the wavefunction formalism compared to the density matrix one. This formalism is implemented by modifying the propagation operator to include non-Hermitian damping and quantum jumps.

With the addition of an absorbing boundary it is possible to treat weakly bound and dissociative vibrational Hamiltonians, whereas via the Monte-Carlo wavefunction method it is feasible to include the effects of the vibrational bath in an inexpensive way. Using these two extensions, we may extend the numerical simulations of linear and nonlinear spectra to a broader range of molecular systems. The reader is encouraged to examine these ideas in the included example MATLAB® code and develop their own simulations exploiting these tools.

Bibliography

[1] Child M S 1991 Analysis of a complex absorbing barrier *Mol. Phys.* **72** 89–3

[2] Dalibard J, Castin Y and Mølmer K 1992 Wave-function approach to dissipative processes in quantum optics *Phys. Rev. Lett.* **68** 580–3

[3] Feit M D, Fleck J A Jr, and Steiger A 1982 Solution of the Schrödinger equation by a spectral method *J. Comp. Phys.* **47** 412–33

[4] Gelin M F, Egorova D and Domcke W 2005 Efficient method for the calculation of time- and frequency-resolved four-wave mixing signals and its application to photon-echo spectroscopy *J. Chem. Phys.* **123** 164112

[5] Heller E J 1981 Frozen Gaussians: a very simple semiclassical approximation *J. Chem. Phys.* **75** 2923–31

[6] Heller E J 1981 The semiclassical way to molecular spectroscopy *Acc. Chem. Res.* **14** 368–75

[7] Heller E J 1991 Cellular dynamics: a new semiclassical approach to time-dependent quantum mechanics *J. Chem. Phys.* **94** 2723–9

[8] Kjellberg P and Pullerits T 2006 Three-pulse photon echo of an excitonic dimer modeled via Redfield theory *J. Chem. Phys.* **124** 024106

[9] Kosloff R and Kosloff D 1986 Absorbing boundaries for wave propagation problems *J. Comp. Phys.* **376** 363–76

[10] Leforestier C, Bisseling R H, Cerjan C, Feit M D, Friesner R, Guldberg A, Hammerich A, Jolicard G, Karrlein W, Meyer H D, Lipkin N, Roncero O and Kosloff R 1991 A comparison of different propagation schemes for the time dependent Schrödinger equation *J. Comp. Phys.* **80** 59–80

[11] Miller W H 2001 The semiclassical initial value representation: a potentially practical way for adding quantum effects to classical molecular dynamics simulations *J. Phys. Chem.* A **105** 2942–55

[12] Muga J G, Palao J P, Navarro B and Egusquiza I L 2004 Complex absorbing potentials *Phys. Rep.* **395** 357–426

[13] Seidner L, Stock G and Domcke W 1995 Nonperturbative approach to femtosecond spectroscopy: General theory and application to multidimensional nonadiabatic photo-isomerization processes *J. Chem. Phys.* **103** 3998–4011

[14] Tannor D J 2007 *Introduction to Quantum Mechanics: A Time Dependent Approach* (Mill Valley, CA: University Science Books)

[15] Yuen-Zhou J, Krich J J and Aspuru-Guzik A 2012 A witness for coherent electronic vs vibronic-only oscillations in ultrafast spectroscopy *J. Chem. Phys.* **136** 234501

Chapter 7

Conclusions

We have provided a pedagogical introduction to nonlinear spectroscopy in light of quantum information processing concepts such as the process matrix $\chi(T)$ and its reconstruction via quantum process tomography. We have built our intuition on spectroscopy by analyzing how energy is exchanged between different light fields interacting with an ensemble of chromophores and have expressed the spectroscopic observables in terms of wavepacket overlaps. Numerical methods and example MATLAB® code to explore these ideas have been provided to help the reader explore complex models that go beyond analytically tractable ones. We have noticed that the spectroscopic measurements give not only signatures, but explicit information about the quantum dynamics triggered in the systems of interest. In particular, we have shown that a series of well-crafted pump–probe type experiments are able to extract the process matrix $\chi(T)$ of the singly-excited manifold of a molecular aggregate. More sophisticated multidimensional spectroscopies can be entirely understood in this framework, and for completeness, we provide a toy calculation on this topic in appendix E. The interested reader can consult [11] to understand how quantum process tomography works in the context of 2D spectroscopy. That work also studies issues of scaling and inversion stability, and provides a comparison of the different nonlinear spectroscopies.

Once we have reconstructed the process matrix $\chi(T)$, which completely characterizes excited-state dynamics, a wealth of questions can be rigorously answered about it. Some examples are: can the vibrational bath $\mathscr{B}$ be described as Markovian? If so, is the secular approximation satisfied, or can a population be transferred to a coherence via the vibrations [4]? If not, what is its degree of non-Markovianity [1, 3, 8]? Does a particular master equation accurately describe the dynamics of the open quantum system? How long does each decoherence process take? How much entanglement is induced in the system upon photoexcitation [10]? Do the vibrations in each site exhibit independent fluctuations or are they somehow correlated [5, 6]? Once these questions

doi:10.1088/978-0-750-31062-8ch7

are answered, interesting questions of control and manipulation of excited states [2, 9] as well as quantum computing with molecular systems [7, 12] can be asked.

Even if full quantum process tomography cannot be performed for a particular system, the rationalization of a nonlinear spectroscopic signal in terms of the process matrix $\chi(T)$ is tremendously valuable, as it indicates the amount of information that is known and unknown about the OQS dynamics, and what is the experimental procedure, if available, to obtain the missing information. The elements of $\chi(T)$, as entries of a large matrix, offer a convenient bookkeeping of this information. We hope to have convinced the reader that the quantum process tomography approach to nonlinear spectroscopy is an intuitive and exciting frontier that will provide us with tools to understand excited-state dynamics more systematically and transparently.

Bibliography

[1] Breuer H P, Laine E M and Piilo J 2009 Measure for the degree of non-Markovian behavior of quantum processes in open systems *Phys. Rev. Lett.* **103** 210401

[2] Brumer P W and Shapiro M 2003 *Principles of the Quantum Control of Molecular Processes* (New York: Wiley-Interscience)

[3] Cheng Y C, Engel G S and Fleming G R 2007 Elucidation of population and coherence dynamics using cross-peaks in two-dimensional electronic spectroscopy *Chem. Phys.* **341** 285–95

[4] Ishizaki A and Fleming G R 2009 On the adequacy of the Redfield equation and related approaches to the study of quantum dynamics in electronic energy transfer *J. Chem. Phys.* **130** 234110

[5] Ishizaki A and Fleming G R 2009 Theoretical examination of quantum coherence in a photosynthetic system at physiological temperature *Proc. Natl Acad. Sci. USA* **106** 17255–60

[6] Kofman A G and Korotkov A N 2009 Two-qubit decoherence mechanisms revealed via quantum process tomography *Phys. Rev.* A **80** 042103

[7] Lozovoy V V and Dantus M 2002 Photon echo pulse sequences with femtosecond shaped laser pulses as a vehicle for molecule-based quantum computation *Chem. Phys. Lett.* **351** 213–21

[8] Rebentrost P and Aspuru-Guzik A 2011 Communication: Exciton–phonon information flow in the energy transfer process of photosynthetic complexes *J. Chem. Phys.* **134** 101103

[9] Rice S and Shao M 2000 *Optical Control of Molecular Dynamics* (New York: Wiley-Interscience)

[10] Sarovar M, Ishizaki A, Fleming G R and Whaley K B 2010 Quantum entanglement in photosynthetic light-harvesting complexes *Nature Phys.* **6** 462–7

[11] Yuen-Zhou J and Aspuru-Guzik A 2011 Quantum process tomography of excitonic dimers from two-dimensional electronic spectroscopy. I. General theory and application to homodimers *J. Chem. Phys.* **134** 134505

[12] Zaari R R and Brown A 2012 Effect of laser pulse shaping parameters on the fidelity of quantum logic gates *J. Chem. Phys.* **137** 104306

IOP Publishing

Ultrafast Spectroscopy

Quantum Information and Wavepackets

Joel Yuen-Zhou *et al*

Appendix A

Mathematical description of a short pulse of light

In this appendix, we will justify the form of the pulses used throughout the book, i.e., equations (3.3) and (3.4). We will consider only one such pulse and assume for simplicity that it propagates along the **z** direction. The generalization to arbitrary numbers of pulses and directions is straightforward.

A plane wave of the form $\boldsymbol{\varepsilon}(\mathbf{r}, t) = \mathrm{e}^{\mathrm{i}(\mathbf{k}\cdot\mathbf{r}-\omega(\mathbf{k})t)}\mathbf{e}_0$ is a solution of Maxwell's equations in the absence of free charges and currents. Here $\mathbf{k}$ is the wavevector of the field, $\omega(\mathbf{k}) = c|\mathbf{k}|$ is the angular frequency at the given $\mathbf{k}$ (c is the speed of light), and $\mathbf{e}_0$ is the polarization. Clearly, this function describes a wave which is delocalized in space and time, since the intensity $|\boldsymbol{\varepsilon}(\mathbf{r},t)|^2$ is constant everywhere. Due to the linearity of Maxwell's equations, we can construct localized solutions by linearly superposing a set of plane waves. Consider for instance,

$$\boldsymbol{\varepsilon}(\mathbf{r}, t) = \int \mathrm{d}^3k A(\mathbf{k} - \mathbf{k}_0)\left(\mathrm{e}^{\mathrm{i}(\mathbf{k}\cdot\mathbf{r}-\omega(\mathbf{k})t)}\mathbf{e}_0 + \mathrm{c.c.}\right), \tag{A.1}$$

where $A(\mathbf{k} - \mathbf{k}_0)$ denotes a distribution of wavevectors centered at $\mathbf{k} = \mathbf{k}_0$. For simplicity, we shall work with a Gaussian distribution of wavevectors centered at $\mathbf{k}_0 = k_0\mathbf{z}$,

$$A(\mathbf{k} - \mathbf{k}_0) = \frac{\eta c \mathrm{e}^{-k_x^2/2\varsigma_{0x}^2}\mathrm{e}^{-k_y^2/2\varsigma_{0y}^2}\mathrm{e}^{-(k_z-k_0)^2/2\varsigma_{0z}^2}\mathrm{e}^{-\mathrm{i}k_z z_0}\mathrm{e}^{\mathrm{i}\phi_0}}{2\pi\sqrt{\left(2\pi\varsigma_{0x}^2\right)\left(2\pi\varsigma_{0y}^2\right)}}, \tag{A.2}$$

with distribution widths ς_{0i} along each dimension i, and $\mathbf{k}$-independent parameters z_0 and ϕ_0, which will be interpreted below. The prefactors have been chosen to yield a convenient result at the end of the calculation.

Assuming narrow distributions along **x** and **y** compared to the one along **z**, $\varsigma_{0x}, \varsigma_{0y} \ll \varsigma_{0z}$, we can approximate $\omega \approx ck_z$, i.e., the plane waves propagate

doi:10.1088/978-0-750-31062-8ch8

mostly along **z**. Inserting equation (A.2) into equation (A.1) yields, after the Fourier transform,

$$\boldsymbol{\varepsilon}(\mathbf{r}, t) = \frac{\eta c \sqrt{2\pi \varsigma_{0z}^2} \mathrm{e}^{-x^2 \varsigma_{0x}^2/2 - y^2 \varsigma_{0y}^2/2 - (z - z_0 - ct)^2 \varsigma_{0z}^2/2} \mathrm{e}^{\mathrm{i}k_0(z - z_o - ct) + \mathrm{i}\phi} \mathbf{e}_0 + \mathrm{c.c.}}{2\pi} \tag{A.3a}$$

$$\approx \frac{\eta c \sqrt{2\pi \varsigma_{0z}^2} \mathrm{e}^{-(z - z_0 - ct)^2 \varsigma_{0z}^2/2} \mathrm{e}^{\mathrm{i}k_0(z - z_0 - ct) + i\phi_0} \mathbf{e}_0 + \mathrm{c.c.}}{2\pi}. \tag{A.3b}$$

Comparing the last two equations, we see that the phase $\mathrm{e}^{\mathrm{i}kz_0}$ localizes the center of the pulse at $z = z_0$ for $t = 0$, and $\mathrm{e}^{\mathrm{i}\phi_0}$ is a global phase offset of the pulse. Equation (A.3) represents a localized Gaussian wavepacket moving at speed c, centered at $z = z_0$ for $t = 0$ and at $z = z_0 + ct$ in general.

In order to simplify equation (A.3), we note that taking $z_0 < 0$ means it takes $t_0 \equiv -\frac{z_0}{c}$ time for the center of the pulse to reach the center of the ensemble, which we assume to be at $z = 0$; this, in turn, gives $\mathrm{e}^{-\mathrm{i}k_0 z_0} = \mathrm{e}^{\mathrm{i}\omega_0 t_0}$. Next, assuming that ς_{0z}^{-1} is much larger than the size of the ensemble of molecules gives $\mathrm{e}^{-(z - z_0 - ct)^2 \varsigma_{0z}^2/2} \approx \mathrm{e}^{-(z_0 - ct)^2 \varsigma_{0z}^2/2}$. This means that molecules located at different positions of the ensemble do not feel radically different *intensities* of the pulse at any given time (but they do experience the spatially varying $\mathrm{e}^{\mathrm{i}k_0 z}$ phase in the *amplitude* of the pulse, which gives rise to phase matching). Finally, we define $\sigma_0 = (c\varsigma_0)^{-1}$ as the temporal duration of the pulse. Altogether, these observations yield,

$$\boldsymbol{\varepsilon}(\mathbf{r}, t) \approx \frac{\eta}{\sqrt{2\pi \sigma_0^2}} \mathrm{e}^{-(t - t_0)/2\sigma_0^2} \mathrm{e}^{\mathrm{i}k_0 z - \mathrm{i}\omega_0(t - t_0) + \mathrm{i}\phi_0} \mathbf{e}_0 + \mathrm{c.c.}, \tag{A.4}$$

which is of the form of equations (3.3) and (3.4).

Appendix B

Validity of time-dependent perturbation theory in the treatment of light–matter interaction

We will now use some typical experimental parameters to verify that the light–matter interaction can be treated perturbatively, as we do throughout the main text. Let us consider some typical values for ultrafast pulses in experiments concerning multi-chromophoric systems, as reported in [2]. In that study, the total energy content of each pulse is $U \approx 500$ pJ, pulses are centered about $\omega_0 \approx 17000\ \text{cm}^{-1}$ and the diameter of the laser spot size is about $2\varsigma_{0x}^{-1} = 2\varsigma_{0y}^{-1} \approx 120\ \mu\text{m}$. This yields a pulse with fluence given by,

$$\begin{aligned}
\text{Fluence} &= \frac{\text{Photons}}{\text{Area}} \\
&\approx \frac{U/\hbar\omega_0}{\pi\frac{1}{\varsigma_x^2}} = \frac{500\times10^{-12}\ \text{J}/(6.626\times10^{-34}\ \text{J s}\times3\times10^{10}\ \text{cm s}^{-1}\times17\times10^{3}\ \text{cm}^{-1})}{\pi(60\times10^{-4}\ \text{cm})^2} \\
&= 1.3\times10^{13}\frac{\text{photons}}{\text{cm}^2}.
\end{aligned} \tag{B.1}$$

We now estimate the value of η from equation (A.3a). The energy content of a pulse is the volume integral of the electric field intensity,

$$\begin{aligned}
U &= \frac{\epsilon_0}{2}\int \mathrm{d}^3\mathbf{r}|\boldsymbol{\varepsilon}(\mathbf{r}, t)|^2 \\
&= \frac{\epsilon_0}{2}\left(\frac{\eta c\varsigma_{0z}}{\sqrt{2\pi}}\right)^2\int \mathrm{d}^3\mathbf{r}e^{-(z-z_0-ct)^2\varsigma_{0z}^2} \\
&= \frac{\epsilon_0 c^2\sqrt{\pi}\eta^2\varsigma_{0z}}{4\varsigma_{0x}\varsigma_{0y}},
\end{aligned} \tag{B.2}$$

where ϵ_0 is the permitivity of free space. Solving for η,

$$\eta = \sqrt{\frac{4\varsigma_{0x}\varsigma_{0y}U}{\epsilon_0 c^2 \varsigma_{0z}\sqrt{\pi}}}$$

$$= \sqrt{\frac{4\varsigma_{0x}\varsigma_{0y}\sigma_0 U}{\epsilon_0 c\sqrt{\pi}}} \tag{B.3}$$

$$\approx \sqrt{\frac{4(1/60\times 10^{-6}\ \mathrm{m})^2(5\times 10^{-15}\ \mathrm{s})(5\times 10^{-10}\ \mathrm{J})}{(8.85\times 10^{-12}\ \mathrm{C^2\,J^{-1}\,m^{-1}})(3\times 10^{8}\ \mathrm{m\,s^{-1}})\sqrt{\pi}}}$$

$$= 7.68\times 10^{-7}\ \mathrm{V\,s\,m^{-1}}. \tag{B.4}$$

Here, we have used (see appendix A) $\sigma_0 = (\varsigma_{0z}c)^{-1} \approx 5$ fs because the pulse under consideration has a power spectrum FWHM of about 1500 cm^{-1}.

Let us now check whether the calculated value for η satisfies the 'weak' strength condition, so that we can confirm the use of time-dependent perturbation theory in our treatment of light–matter interaction. From the theory of two-level systems [1], one knows that the rate of transfer of amplitude transfer between the two states driven by an ac-field is given by the Rabi frequency, which in resonance reads as $|\Omega_0| = |\boldsymbol{\mu}\cdot \mathbf{e}_0\varepsilon_0(t-t_0)| \leqslant |\mu\eta|\sqrt{2\pi\sigma_0^2}$. We take μ to be a characteristic value of the electronic transition dipole moment for a chromophore, $\mu \approx 5$ D. For a pulse to be weak compared to $H_0(\boldsymbol{R})$ (equation (2.1)) means that it will not be strong enough to induce a full Rabi cycle between the two states while it is on. This amounts to saying that the time for one cycle to occur will be much larger than some characteristic time involving the width of the pulse (in this case, let us take this to be $3\sigma_0$),

$$\text{Period of a Rabi cycle} \gg 3\sigma_0$$

$$\frac{2\pi\hbar}{\left(\mu\frac{c\varsigma_0}{\sqrt{2\pi}}\eta\right)} \gg \frac{3}{c\varsigma_0}, \tag{B.5}$$

or alternatively,

$$\eta \ll \frac{(2\pi)^{3/2}\hbar}{3\mu}$$

$$\approx \frac{(2\pi)^{3/2}(1.05\times 10^{-34}\ \mathrm{J\,s})}{3(3.3\times 10^{-29}\ \mathrm{C\,m})}$$

$$= 3.34\times 10^{-5}\ \mathrm{V\,s\,m^{-1}}. \tag{B.6}$$

By comparing equation (B.4) with equation (B.6), we justify the use of time-dependent perturbation theory for the description of the mentioned ultrafast spectroscopy experiments.

Bibliography

[1] Allen L and Eberly J H 1987 *Optical Resonance and Two-Level Atoms* (New York: Dover Publications)

[2] Womick J M, Miller S A and Moran A M 2009 Correlated exciton fluctuations in cylindrical molecular aggregates *J. Phys. Chem.* B **113** 6630–9

Appendix C

Many-molecule quantum states of an ensemble of chromophores interacting with coherent light

Throughout the book, we have addressed spectroscopies associated with an ensemble of molecules that do not interact with each other but couple to the same coherent light fields. In doing so, we have deemed it sufficient to consider the effects of the optical perturbations on each molecule independently and then establish phase-matching by keeping track of the spatially dependent phases $e^{i\mathbf{k}_n\cdot\mathbf{r}+i\phi_n}$ at the end of the calculation. A question arises: Are we missing physically relevant processes by using this 'independent-molecule' procedure and should we instead consider a 'many-molecule' calculation of the quantum state of the ensemble? Our goal in this appendix is to prove that the many-molecule approach yields the same result as the independent-molecule one, rendering the methods in the book valid. To keep the notation simple, we shall work within the wave-function formalism, although extension of the conclusions to density matrices are straightforward.

Let us redo the calculation of the polarization induced by light in an ensemble of identical chromophores (for instance, but not restricted to, coupled dimers) in the PP' setup. This time, we shall keep track of the evolving many-molecule state of the ensemble. The latter consists of N noninteracting molecules, so it admits a wave-function description in the product form

$$|\Upsilon(t)\rangle = |\Psi(\mathbf{r}_1, t)\rangle|\Psi(\mathbf{r}_2, t)\rangle\cdots|\Psi(\mathbf{r}_N, t)\rangle, \tag{C.1}$$

where $|\Psi(\mathbf{r}_j, t)\rangle$ describes the electronic quantum state of the jth molecule centered at $\mathbf{r}_j$. The spatial phases due to the fields are all contained in $|\Upsilon(t)\rangle$. In the perturbative regime, each $|\Psi(\mathbf{r}_i,t)\rangle$ takes an analogous form to equation (3.15). Just as $|\Psi(\mathbf{r}_j,t)\rangle$ can be expanded in various powers of the field strength λ, so can $|\Upsilon(t)\rangle$. Enumerating only some of the perturbative wavefunctions resulting from the various actions of P and P', $|\Upsilon(t)\rangle$ acquires the form

doi:10.1088/978-0-750-31062-8ch10

$$
\begin{aligned}
|\Upsilon(t)\rangle &= |\Upsilon_0(t)\rangle + |\Upsilon_{+P}(t)\rangle + |\Upsilon_{+P'}(t)\rangle \\
&\quad + |\Upsilon_{+P-P}(t)\rangle + |\Upsilon_{+P-P'}(t)\rangle + |\Upsilon_{+P+P'}(t)\rangle \\
&\quad + |\Upsilon_{+P-P+P'}(t)\rangle + \cdots
\end{aligned} \tag{C.2}
$$

Here, we have followed the notation of section 3.2, where the string of subscripts indicates the time-ordering of the pulse actions. For example, $|\Upsilon_{+P-P+P'}(t)\rangle$ is the perturbative wavefunction resulting from the action on $|\Upsilon_0(t)\rangle$ due to ε_P first, ε_P^* next, and $\varepsilon_{P'}$ last. We are interested in computing the polarization $\mathbf{P}(\mathbf{r}_j, t) = \langle\Upsilon(t)|\boldsymbol{\mu}^{(j)}|\Upsilon(t)\rangle$ at an arbitrary molecular location $\mathbf{r}_j$ within the ensemble, where $\boldsymbol{\mu}^{(j)}$ is the same dipole operator as in equation (2.9), but we have emphasized its locality with a subscript that indicates that it acts exclusively on molecule j.

Starting at zeroth-order, we get the trivial tensor product result,

$$
|\Upsilon_0(t)\rangle = |\Psi_0(t)\rangle|\Psi_0(t)\rangle|\Psi_0(t)\rangle, \tag{C.3}
$$

where each of the $|\Psi_0(t)\rangle$ is given by equation (3.18a). At first order, we obtain

$$
\begin{aligned}
|\Upsilon_{+n}(t)\rangle &= \left(\mathrm{e}^{ik_n\cdot\mathbf{r}_1+\phi_n}|\Psi_{+n}(t)\rangle\right)|\Psi_0(t)\rangle\cdots|\Psi_0(t)\rangle \\
&\quad + |\Psi_0(t)\rangle\left(\mathrm{e}^{ik_n\cdot\mathbf{r}_2+\phi_n}|\Psi_{+n}(t)\rangle\right)\cdots|\Psi_0(t)\rangle \\
&\quad + \cdots \\
&\quad + |\Psi_0(t)\rangle|\Psi_0(t)\rangle\cdots\left(\mathrm{e}^{ik_n\cdot\mathbf{r}_N+\phi_n}|\Psi_{+n}(t)\rangle\right) \\
&= \sum_j \mathrm{e}^{ik_n\cdot\mathbf{r}_j+\phi_n}|\Psi_0(t)\rangle\cdots\underbrace{|\Psi_{+n}(t)\rangle}_{j\text{th molecule}}\cdots|\Psi_0(t)\rangle.
\end{aligned} \tag{C.4}
$$

That is, $|\Upsilon_{+n}(t)\rangle$ is a *coherent* superposition of N terms corresponding to all possible states with a single molecule in the SEM and the rest in the GSM. Computing one of the polarization terms to first order in λ, we get

$$
\langle\Upsilon_0(t)|\boldsymbol{\mu}^{(j)}|\Upsilon_{+P'}(t)\rangle = \mathrm{e}^{\mathrm{i}\mathbf{k}_{P'}\cdot\mathbf{r}+\mathrm{i}\phi_{P'}}\underbrace{\langle\Psi_0(t)|\boldsymbol{\mu}^{(j)}|\Psi_{+P'}(t)\rangle}_{=\mathbf{P}^{(1)}_{\mathbf{k}_{P'}}(t)}, \tag{C.5}
$$

which validates the independent-molecule calculation of linear absorption in equation (3.32).

At second order, one of the possible wavefunctions is

$$
\begin{aligned}
|\Upsilon_{+P+P'}(t)\rangle &= \sum_j \mathrm{e}^{\mathrm{i}\mathbf{k}_P\cdot\mathbf{r}_j+\mathrm{i}\mathbf{k}_{P'}\cdot\mathbf{r}_j+\phi_P+\phi_{P'}}|\Psi_0(t)\rangle\cdots\underbrace{|\Psi_{+P+P'}(t)\rangle}_{j\text{th molecule}}\cdots|\Psi_0(t)\rangle \\
&\quad + \sum_{j,l\neq j} \mathrm{e}^{\mathrm{i}\mathbf{k}_P\cdot\mathbf{r}_j+\mathrm{i}\mathbf{k}_{P'}\cdot\mathbf{r}_l+\phi_P+\phi_{P'}}|\Psi_0(t)\rangle\cdots\underbrace{|\Psi_{+P}(t)\rangle}_{j\text{th molecule}}\cdots\underbrace{|\Psi_{+P'}(t)\rangle}_{l\text{th molecule}}\cdots|\Psi_0(t)\rangle.
\end{aligned} \tag{C.6}
$$

The rest of the second-order wavefunctions have a similar structure but different subscripts. For the moment, let us concentrate on equation (C.6). $|\Upsilon_{+P+P'}(t)\rangle$ is a coherent superposition of two types of terms. The first class refers to N states with one molecule in the DEM and the rest in the GSM. The second class consists of $N(N-1)$ states with two molecules in the SEM and the rest in the GSM. So far, the calculations in the book ignore the second class. However, for large N, they are the largest contribution to $|\Upsilon_{+P+P'}(t)\rangle$. Are we in trouble?

As an illustration, let us generalize the PP' polarization calculated in equation (5.34) to the many-molecule case. Computing such polarization at the jth molecule,

$$\begin{aligned}
&e^{i\mathbf{k}_{P'}\cdot\mathbf{r}_j}\mathbf{P}^{(3)\,\text{many molecule}}_{\mathbf{k}_P}(t)\\
&\quad\equiv \underbrace{\langle\Upsilon_{+P}(t)|\boldsymbol{\mu}^{(j)}|\Upsilon_{+P+P'}(t)\rangle}_{\text{ESA}} + \underbrace{\langle\Upsilon_{+P-P'}(t)|\boldsymbol{\mu}^{(j)}|\Upsilon_{+P}(t)\rangle}_{\text{SE}}\\
&\quad + \underbrace{\langle\Upsilon_{+P-P}(t)|\boldsymbol{\mu}^{(j)}|\Upsilon_{+P'}(t)\rangle + \langle\Upsilon_{0}(t)|\boldsymbol{\mu}^{(j)}|\Upsilon_{+P-P+P'}(t)\rangle}_{\text{GSB}},
\end{aligned}\tag{C.7}$$

where we have labeled each term by the physical process it represents. Each of the wavefunctions appearing in this equation is of the form or a variant of equations (C.3), (C.4), or (C.6). Furthermore, we have factored out the spatial phase $e^{i\mathbf{k}_P\cdot\mathbf{r}_j}$ from the definition of $\mathbf{P}^{(3)\,\text{many molecule}}_{\mathbf{k}_P}(t)$ to make a direct analogy with $\mathbf{P}^{(3)}_{\mathbf{k}_P}(t)$ (see equation (3.33)) afterwards.

Listing the various resulting terms,

$$\begin{aligned}
&\langle\Upsilon_{+P}(t)|\boldsymbol{\mu}^{(j)}|\Upsilon_{+P+P'}(t)\rangle\\
&\quad=\underbrace{\sum_{i,j',l} e^{-i\mathbf{k}_P\cdot\mathbf{r}_i+i\mathbf{k}_P\cdot\mathbf{r}_{j'}+i\mathbf{k}_{P'}\cdot\mathbf{r}_l}\langle\Psi_{+P}(t)|\boldsymbol{\mu}^{(j)}|\Psi_{+P+P'}(t)\rangle\delta_{ij}\delta_{j'j}\delta_{lj}}_{\equiv\mathscr{X}_{\text{ESA}}}\\
&\quad+\underbrace{\sum_{i,j',l\neq i} e^{-i\mathbf{k}_P\cdot\mathbf{r}_i+i\mathbf{k}_P\cdot\mathbf{r}_{j'}+i\mathbf{k}_{P'}\cdot\mathbf{r}_l}\left(\langle\Psi_{0}(t)|\boldsymbol{\mu}^{(j)}|\Psi_{+P'}(t)\rangle\delta_{lj}\right)\left(\langle\Psi_{+P}(t)|\Psi_{+P}(t)\rangle\delta_{ij'}\right)}_{\equiv\mathscr{Y}_{\text{ESA},1}}\\
&\quad+\underbrace{\sum_{i,j'\neq i,l} e^{-i\mathbf{k}_P\cdot\mathbf{r}_i+i\mathbf{k}_P\cdot\mathbf{r}_{j'}+i\mathbf{k}_{P'}\cdot\mathbf{r}_l}\left(\langle\Psi_{0}(t)|\boldsymbol{\mu}^{(j)}|\Psi_{+P}(t)\rangle\delta_{j'j}\right)\left(\langle\Psi_{+P}(t)|\Psi_{+P'}(t)\rangle\delta_{il}\right)}_{\equiv\mathscr{Y}_{\text{ESA},2}}.\\
&\langle\Upsilon_{+P-P'}(t)|\boldsymbol{\mu}^{(j)}|\Upsilon_{+P}(t)\rangle\\
&\quad=\underbrace{\sum_{i,j',l} e^{-i\mathbf{k}_P\cdot\mathbf{r}_i+i\mathbf{k}_P\cdot\mathbf{r}_{j'}+i\mathbf{k}_{P'}\cdot\mathbf{r}_l}\left(\langle\Psi_{+P-P'}(t)|\boldsymbol{\mu}^{(j)}|\Psi_{+P}(t)\rangle\delta_{ij}\delta_{j'j}\delta_{lj}\right)}_{\equiv\mathscr{X}_{\text{SE}}}\\
&\quad+\underbrace{\sum_{i,j'\neq i,l} e^{-i\mathbf{k}_P\cdot\mathbf{r}_i+i\mathbf{k}_P\cdot\mathbf{r}_{j'}+i\mathbf{k}_{P'}\cdot\mathbf{r}_l}\left(\langle\Psi_{0}(t)|\boldsymbol{\mu}^{(j)}|\Psi_{+P}(t)\rangle\delta_{j'j}\right)\left(\langle\Psi_{+P-P'}(t)|\Psi_{0}(t)\rangle\delta_{il}\right)}_{\equiv\mathscr{Y}_{\text{SE}}},
\end{aligned}\tag{C.8}$$

$$\langle \Upsilon_{+P-P}(t)|\boldsymbol{\mu}^{(j)}|\Upsilon_{+P'}(t)\rangle = \underbrace{\sum_{i,j',l} \mathrm{e}^{-\mathrm{i}\mathbf{k}_P\cdot\mathbf{r}_i+\mathrm{i}\mathbf{k}_P\cdot\mathbf{r}_{j'}+\mathrm{i}\mathbf{k}_{P'}\cdot\mathbf{r}_l}\left(\langle \Psi_{+P-P'}(t)|\boldsymbol{\mu}^{(j)}|\Psi_{+P}(t)\rangle\delta_{ij}\delta_{j'j}\delta_{lj}\right)}_{\equiv\mathscr{X}_{\mathrm{GSB},1}} + \underbrace{\sum_{i,j',l\neq i} \mathrm{e}^{-\mathrm{i}\mathbf{k}_P\cdot\mathbf{r}_i+\mathrm{i}\mathbf{k}_P\cdot\mathbf{r}_{j'}+\mathrm{i}\mathbf{k}_{P'}\cdot\mathbf{r}_l}\left(\langle \Psi_0(t)|\boldsymbol{\mu}^{(j)}|\Psi_{+P'}(t)\rangle\delta_{lj}\right)\left(\langle \Psi_{+P-P}(t)|\Psi_0(t)\rangle\delta_{ij'}\right)}_{\equiv\mathscr{Y}_{\mathrm{GSB},1}}, \tag{C.9}$$

$$\langle \Upsilon_0(t)|\boldsymbol{\mu}^{(j)}|\Upsilon_{+P-P+P'}(t)\rangle = \underbrace{\sum_{i,j',l} \mathrm{e}^{-\mathrm{i}\mathbf{k}_P\cdot\mathbf{r}_i+\mathrm{i}\mathbf{k}_P\cdot\mathbf{r}_{j'}+\mathrm{i}\mathbf{k}_{P'}\cdot\mathbf{r}_l}\left(\langle \Psi_0(t)|\boldsymbol{\mu}^{(j)}|\Psi_{+P-P+P'}(t)\rangle\delta_{ij}\delta_{j'j}\delta_{lj}\right)}_{\equiv\mathscr{X}_{\mathrm{GSB},2}} + \underbrace{\sum_{i,j',l\neq i} \mathrm{e}^{-\mathrm{i}\mathbf{k}_P\cdot\mathbf{r}_i+\mathrm{i}\mathbf{k}_P\cdot\mathbf{r}_{j'}+\mathrm{i}\mathbf{k}_{P'}\cdot\mathbf{r}_l}\left(\langle \Psi_0(t)|\boldsymbol{\mu}^{(j)}|\Psi_{+P'}(t)\rangle\delta_{lj}\right)\left(\langle \Psi_0(t)|\Psi_{+P-P}(t)\rangle\delta_{ij'}\right)}_{\equiv\mathscr{Y}_{\mathrm{GSB},2}}. \tag{C.10}$$

Although lengthy, these expressions are easily interpreted. Let us first consider the sum of $\mathscr{X}$ terms, associated with actions of every pulse in molecule j,

$$\begin{aligned} \mathscr{X}_{\mathrm{ESA}} + \mathscr{X}_{\mathrm{SE}} + \mathscr{X}_{\mathrm{GSB},1} + \mathscr{X}_{\mathrm{GSB},2} &= \mathrm{e}^{\mathrm{i}\mathbf{k}_{P'}\cdot\mathbf{r}_j}\Big(\langle \Psi_{+P}(t)|\boldsymbol{\mu}^{(j)}|\Psi_{+P+P'}(t)\rangle + \langle \Psi_{+P-P'}(t)|\boldsymbol{\mu}^{(j)}|\Psi_{+P}(t)\rangle \qquad \text{(C.11)} \\ &\quad + \langle \Psi_{+P-P}(t)|\boldsymbol{\mu}^{(j)}|\Psi_{+P'}(t)\rangle + \langle \Psi_0(t)|\boldsymbol{\mu}^{(j)}|\Psi_{+P-P+P'}(t)\rangle\Big). \\ &= \mathrm{e}^{\mathrm{i}\mathbf{k}_{P'}\cdot\mathbf{r}_j}\mathbf{P}^{(3)}_{\mathbf{k}_P} \qquad \text{(C.12)} \end{aligned}$$

where the last line relates this sum to the total contribution to the PP' polarization that we expect from the single-molecule calculation in equation (3.33).

The $\mathscr{Y}$ terms do not appear in any of our calculations in the book. If the methods presented in the book are to be valid, they ought to cancel. Let us understand what sort of physical processes they describe. Looking at the ESA terms first, $\mathscr{Y}_{\mathrm{ESA},1}$ is the ESA polarization caused by linear absorption of pulse P' in molecule j in the presence of an excited-state population $\langle \Psi_{+P}(t)|\Psi_{+P}(t)\rangle$ at *any* other molecule $i \neq j$ (hence, the sum over all possible i indices). $\mathscr{Y}_{\mathrm{ESA},2}$ is analogous to $\mathscr{Y}_{\mathrm{ESA},1}$, but refers instead to the linear absorption of P by molecule j given an excited-state overlap $\langle \Psi_{+P}(t)|\Psi_{+P'}(t)\rangle$ in a different molecule. It is clear that the $\mathscr{Y}$ terms correspond to nonlinear polarization contributions associated with photoexcitation processes in different noninteracting molecules across the ensemble. We now show that these ESA terms, corresponding to energy lost by the probe pulse P', precisely cancel the corresponding SE and GSB terms, related to energy gained by it (see figure C.1). In order to see this, we note that

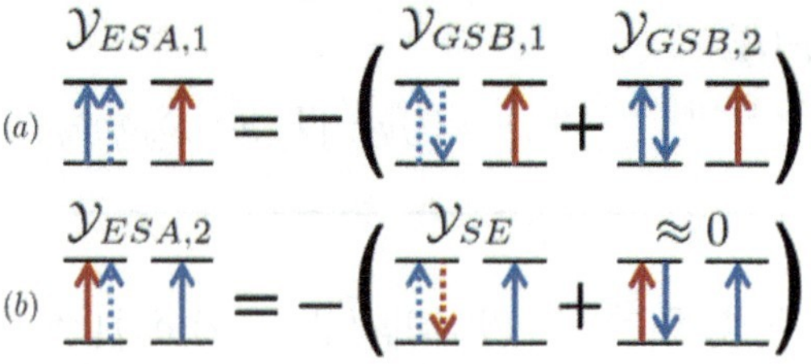

Figure C.1. Third-order polarization pathways involving two independent chromophores interacting with the same coherent light pulses. Arrows indicate actions of pump P (blue) and probe P' (red) in bra (dashed arrows) and ket (solid arrows). (*a*) Pathway where P' excites right chromophore in the presence of excited-state population prepared by P in left chromophore. This ESA contribution $\mathscr{Y}_{ESA,1}$ exactly cancels two GSB processes $\mathscr{Y}_{GSB,1}$ and $\mathscr{Y}_{GSB,2}$ associated with left chromophore first excited and subsequently de-excited by P. (*b*) A similar pathway where instead P excites right chromophore in the presence of excited-state overlap prepared by P and P' in left chromophore. Now, this ESA contribution $\mathscr{Y}_{ESA,2}$ approximately cancels a SE process $\mathscr{Y}_{SE}$ where where left chromophore gets excited first by P and subsequently de-excited by P'. The rest of the cancellation is given by a negligible contribution where the action of the pulses is reversed.

$$
\begin{aligned}
&\langle\Psi_{+P-P}(t)|\Psi_0(t)\rangle + \langle\Psi_0(t)|\Psi_{+P-P}(t)\rangle \\
&= \int_{-\infty}^{t} dt' \int_{-\infty}^{t'} dt'' \langle\Psi_0(t'')|\{-i\boldsymbol{\mu}\cdot\mathbf{e}_P^*\varepsilon_P^*(t''-t_P)\}e^{-iH_0(t''-t')} \\
&\quad \times\{-i\boldsymbol{\mu}\cdot\mathbf{e}_P\varepsilon_P(t'-t_P)\}e^{-iH_0(t'-t)}|\Psi_0(t)\rangle \\
&\quad + \int_{-\infty}^{t} dt'' \int_{-\infty}^{t''} dt' \langle\Psi_0(t)|e^{-iH_0(t-t'')}\{i\boldsymbol{\mu}\cdot\mathbf{e}_P^*\varepsilon_P^*(t''-t_P)\}e^{-iH_0(t''-t')} \\
&\quad \times\{i\boldsymbol{\mu}\cdot\mathbf{e}_P\varepsilon_P(t'-t_P)\}|\Psi_0(t')\rangle \\
&= -\underbrace{\int_{-\infty}^{t} dt'' \langle\Psi_0(t'')|\{-i\boldsymbol{\mu}\cdot\mathbf{e}_P^*\varepsilon_P^*(t''-t_P)\}e^{-iH_0(t''-t)}}_{=\langle\Psi_{+P}(t)|} \\
&\quad \times \underbrace{\int_{-\infty}^{t} dt' e^{-iH_0(t-t')}\{i\boldsymbol{\mu}\cdot\mathbf{e}_P\varepsilon_P(t'-t_P)\}|\Psi_0(t')\rangle}_{=|\Psi_{+P'}(t)\rangle}. \\
&= -\langle\Psi_{+P}(t)|\Psi_{+P}(t)\rangle,
\end{aligned}
\tag{C.13}
$$

which imply $\mathscr{Y}_{\text{ESA},1} = -\mathscr{Y}_{\text{GSB},1} - \mathscr{Y}_{\text{GSB},2}$, and similarly,

$$
\begin{aligned}
\langle\Psi_{+P-P'}(t)|\Psi_0(t)\rangle &\approx \langle\Psi_{+P-P'}(t)|\Psi_0(t)\rangle + \underbrace{\langle\Psi_0(t)|\Psi_{+P'-P}(t)\rangle}_{\approx 0} \\
&= -\langle\Psi_{+P}(t)|\Psi_{+P'}(t)\rangle,
\end{aligned}
\tag{C.14}
$$

which yields $\mathscr{Y}_{\text{ESA},2} \approx -\mathscr{Y}_{\text{SE}}$. The reason why this last identity is just approximate and not exact is because equation (C.7) has implicitly assumed that $t_P \ll t_{P'}$ and ignored terms where P' acts before P (i.e., $|\Psi_{+P'-P}(t)\rangle \approx 0$), which in itself is an approximation. Upon inclusion in the polarization of every term beyond the RWA as well as every pulse action time-ordering, the cancellation of terms which involve two or more noninteracting chromophores becomes formally exact. Hence, it follows that equation (C.11) is the only contribution to equation (C.7), and $\mathbf{P}^{(3)\text{ many molecule}}_{\mathbf{k}_{P'}} = \mathbf{P}^{(3)}_{\mathbf{k}_{P'}}$ and the single-molecule method in the book is justified.

This argument can be generalized to arbitrary phase-matching setups, where the relevant generalization of equation (C.13) is,

$$\langle \Psi_{(\text{string a})+n+n'}(t)|\Psi_{(\text{string b})}(t)\rangle + \langle \Psi_{(\text{string a})}(t)|\Psi_{(\text{string b})-n'-n}(t)\rangle$$
$$= -\langle \Psi_{(\text{string a})+n}(t)|\Psi_{(\text{string b})-n'}(t)\rangle.$$

Here, 'string a' and 'string b' refer to arbitrary strings of pulse actions. The intuition for the corresponding cancellation is still the same: for every ESA process involving two or more noninteracting chromophores, there are corresponding SE and GSB processes that compensate for it. Note that this result holds for ensembles of molecules interacting with coherent light fields. It is intriguing to investigate whether it holds for more general fields, e.g., entangled photons, which can induce correlations between molecules even when they do not interact with each other in the absence of the field (see for instance, [1]). This is a subject of ongoing investigation.

Bibliography

[1] Ashok Muthukrishnan A, Agarwal G S and Scully M O 2004 Inducing disallowed two-atom transitions with temporally entangled phonons *Phys. Rev. Lett.* **93** 093002

IOP Publishing

Ultrafast Spectroscopy
Quantum Information and Wavepackets
Joel Yuen-Zhou *et al*

Appendix D

Frequency-resolved spectroscopy

With the exception of section 6.3, this book considers spectroscopic signals that result from a frequency-integrated count of photons. When a spectrophotometer is available, more information can be obtained by performing the count by photon frequency. Whenever the LO satisfies the phase-matching condition with a particular polarization component $\mathbf{P}_{\mathbf{k}_s}$ (equation (3.50)), the absorption spectrum at frequency ω in the direction k_s, called $A_{\mathbf{k}_s}(\omega)$, is defined as the ratio between the reduction in LO energy per unit area absorbed by the sample at a particular frequency (equation (3.41)) and the photon frequency ω,

$$\sum_r \Delta E_{\mathrm{LO}}(\mathbf{r},\omega) = \frac{N}{V}(2\pi)^3\delta^3(\mathbf{k}_{\mathrm{LO}}-\mathbf{k}_s)\left(\frac{|\tilde{\varepsilon}_{\mathrm{LO}}(\omega)|^2 c}{4\pi}\right)A_{\mathbf{k}_s}(\omega) \tag{D.1}$$

where c is the speed of light in the medium where the molecules reside, r denotes the molecular positions in the ensemble, which, for simplicity, are assumed to be uniformly distributed in such a medium, N is the number of molecules in the ensemble and V is the volume that they occupy. By using equations (3.41), (3.47), and carrying out the sum of equation (D.1) with the identity in equation (3.50), we can solve for $A_{\mathbf{k}_s}(\omega)$ to obtain,

$$\boxed{A_{\mathbf{k}_s}(\omega) = \frac{4\pi\omega}{|\tilde{\varepsilon}_{\mathrm{LO}}(\omega)|^2 c}\Im \mathrm{e}^{\mathrm{i}(\phi_s-\phi_{\mathrm{LO}})}\mathrm{e}^{-\mathrm{i}\omega t_{\mathrm{LO}}}\tilde{\varepsilon}^*_{\mathrm{LO}}(\omega)\mathbf{e}^*_{\mathrm{LO}}\cdot\tilde{\mathbf{P}}_{\mathbf{k}_s}(\omega).} \tag{D.2}$$

To keep the exposition simple, we shall only develop two examples of the use of equation (D.2). One will be on the most common type of absorption spectra, linear absorption, and the other will be on frequency-resolved PP' spectroscopy.

Spectrally resolving a signal adds an extra frequency dimension that corresponds, in the Fourier transform sense, to another time-interval. Hence, there is a *frequency-integrated* signal which gives analogous information to that given by the *frequency-resolved* $A_{\mathbf{k}_s}(\omega)$. As an example, as we shall see, the frequency-resolved PP' spectrum gives analogous information to a frequency-integrated TG signal. In fact, there exists a plethora of combinations of time and frequency-resolved or frequency-integrated spectra which can be harnessed to yield identical or similar quantum dynamical information about a specific system. The reader is invited to explore these opportunities, which are subjects of present research.

Example 12. Time and frequency domain pictures of linear absorption spectroscopy

Consider the linear absorption experiment described in section 4.1.

1. Derive an expression for the linear absorption spectrum $A_{\mathbf{k}_P}(\omega)$ in terms of the Fourier transform of $\langle\Psi_P|\mathrm{e}^{-\mathrm{i}H_0t}|\Psi_P\rangle$.
2. Derive another expression for $A_{\mathbf{k}_P}(\omega)$ in terms of the dipole–dipole correlation function, $\langle\Psi_0|(\boldsymbol{\mu}\cdot\mathbf{e}_P^*)\mathrm{e}^{-\mathrm{i}H_0t}(\boldsymbol{\mu}\cdot\mathbf{e}_\mathbf{P})|\Psi_0\rangle$, and compare it with the answer to part 1.
3. Evaluate the result from part 2 in a complete set of vibronic eigenstates $H_0|m\rangle = E_m|m\rangle$ and comment on the use of this frequency-domain formula as opposed to the other ones in the time domain.

Solution

1. The absorption spectrum in equation (D.2) is of the form $A_{\mathbf{k}_s}(\omega) \propto \tilde{f}^*(\omega)\tilde{g}(\omega)$, where $f = \varepsilon$ and $g = \mathbf{P}$. A product in the frequency domain transforms into a convolution in the time domain. Following the convention from equation (3.42),

$$\tilde{f}(\omega)\tilde{g}(\omega) = \int_{-\infty}^{\infty} \mathrm{d}t\, \mathrm{e}^{\mathrm{i}\omega t}\left[\int_{-\infty}^{\infty} \mathrm{d}t' f(t-t')g(t')\right], \tag{D.3a}$$

$$\tilde{f}^*(\omega)\tilde{g}(\omega) = \int_{-\infty}^{\infty} \mathrm{d}t\, \mathrm{e}^{\mathrm{i}\omega t}\left[\int_{-\infty}^{\infty} \mathrm{d}t' f^*(t'-t)g(t')\right]. \tag{D.3b}$$

As explained in section 4.1, in the linear-absorption experiment, pulse P can be regarded both as the trigger of the polarization in the direction $\mathbf{k}_P$ as well as the LO. Here, $t_{\mathrm{LO}} = t_P$ and $\phi_s = \phi_{\mathrm{LO}} = \phi_P$, so the phases do not play a role in the absorption spectrum of equation (D.2). The frequency-domain expressions for the pulse $f(t) = \varepsilon_P(t-t_P)\mathbf{e}_P$ and the polarization $\mathbf{P}_{\mathbf{k}_P} = \langle\Psi_0(t)|\boldsymbol{\mu}|\Psi_P(t)\rangle$ are $\tilde{f}(\omega) = \mathrm{e}^{\mathrm{i}\omega t_P}\tilde{\varepsilon}_P(\omega)\mathbf{e}_P$ (see equation (3.44)) and $\tilde{g}(\omega) = \tilde{\mathbf{P}}_{\mathbf{k}_P}(\omega) = \int_{-\infty}^{\infty}\mathrm{d}t\,\mathrm{e}^{\mathrm{i}\omega t}\mathbf{P}_{\mathbf{k}_P}(t)$, respectively. Using the definition of $|\Psi_P(t)\rangle$ in equation (4.2), the convolution corresponding to equation (D.2) can be written as,

$$\begin{aligned}
\int_{-\infty}^{\infty} \mathrm{d}t' f^*(t'-t)g(t') &= \int_{-\infty}^{\infty} \mathrm{d}t'\varepsilon_P^*(t'-t-t_P)\langle\Psi_0(t')|\boldsymbol{\mu}\cdot\mathbf{e}_P^*|\Psi_P(t')\rangle \\
&= \mathrm{i}\int_{-\infty}^{\infty} \mathrm{d}\tau\left[\langle\Psi_0(\tau)|\mathrm{e}^{\mathrm{i}E_0t}(-\mathrm{i})\{\varepsilon_P^*(\tau-t_P)\boldsymbol{\mu}\cdot\mathbf{e}_P^*\}\right]\mathrm{e}^{-\mathrm{i}H_0t}|\Psi_P(\tau)\rangle \\
&= \mathrm{i}\mathrm{e}^{\mathrm{i}E_0t}\int_{-\infty}^{\infty} \mathrm{d}\tau[\partial_\tau\langle\Psi_P(\tau)|]\mathrm{e}^{-\mathrm{i}H_0t}|\Psi_P(\tau)\rangle
\end{aligned} \tag{D.4}$$

In the second line, we have changed variables from $t' \to \tau = t' - t$, and recognized that $|\Psi_0(t')\rangle = \mathrm{e}^{-\mathrm{i}E_0 t}|\Psi_0(\tau)\rangle$ since $|\Psi_0\rangle$ is an eigenstate of H_0. We have then proceeded as in equation (4.5), by identifying the bra as $\partial_\tau\langle\Psi_P(\tau)|$. Fourier transforming the result of equation (D.4) as in (D.3*b*) and adapting it to (D.2) yields

$$\begin{aligned}
A_{k_P}(\omega) &= \frac{4\pi\omega}{|\tilde{\varepsilon}_P(\omega)^2|c}\Im\int_{-\infty}^{\infty} \mathrm{d}t\mathrm{e}^{\mathrm{i}\omega t}(\mathrm{i}\mathrm{e}^{\mathrm{i}E_0 \mathrm{t}})\int_{-\infty}^{\infty} \mathrm{d}\tau[\partial_\tau\langle\Psi_P(\tau)|]\mathrm{e}^{-\mathrm{i}H_0 t}|\Psi_P(\tau)\rangle \\
&= \frac{4\pi\omega}{|\tilde{\varepsilon}_P(\omega)^2|c}\Re\int_{-\infty}^{\infty} \mathrm{d}t\mathrm{e}^{\mathrm{i}(\omega+E_0)\mathrm{t}}\int_{-\infty}^{\infty} \mathrm{d}\tau[\partial_\tau\langle\Psi_P(\tau)|]\mathrm{e}^{-\mathrm{i}H_0 t}|\Psi_P(\tau)\rangle \\
&= \frac{2\pi\omega}{|\tilde{\varepsilon}_P(\omega)^2|c}\int_{-\infty}^{\infty} \mathrm{d}t\mathrm{e}^{\mathrm{i}(\omega+E_0)\mathrm{t}}\int_{-\infty}^{\infty} \mathrm{d}\tau[\partial_\tau\langle\Psi_P(\tau)|]\mathrm{e}^{-\mathrm{i}H_0 t}|\Psi_P(\tau)\rangle \\
&+ \langle\Psi_P(\tau)|\mathrm{e}^{-\mathrm{i}H_0 t}[\partial_\tau|\Psi_P(\tau)\rangle] \\
&= \frac{2\pi\omega}{|\tilde{\varepsilon}_P(\omega)^2|c}\int_{-\infty}^{\infty} \mathrm{d}t\mathrm{e}^{\mathrm{i}(\omega+E_0)\mathrm{t}}\langle\Psi_P|\mathrm{e}^{-\mathrm{i}H_0 t}|\Psi_P\rangle. \qquad \text{(D.5)}
\end{aligned}$$

To get from the third to the fourth line, we have used the definition of the asymptotic wavepacket in equation (4.3) (similar derivations were carried out for equations (4.34)–(4.36)). Equation (D.5) has a very intuitive appeal. Once the pulse P is fully absorbed, an $O(\eta)$ asymptotic wavepacket is created (equation (4.3)). This wavepacket then evolves for a time t via $\mathrm{e}^{-\mathrm{i}H_0 t}$, and the absorption spectrum depends on its overlap with the original instance at $t = 0$. As a simple check, let us divide equation (D.5) by ω and integrate across all frequencies to obtain the total number of absorbed photons,

$$\int_{-\infty}^{\infty} \mathrm{d}\omega \frac{|\tilde{\varepsilon}_P(\omega)|^2 A_{\mathbf{k}_P}(\omega)}{\omega} = \frac{4\pi^2\langle\Psi_P|\Psi_P\rangle}{c} \propto S_P, \qquad \text{(D.6)}$$

which confirms equation (4.5).

2. Equation (D.5) is not written in the most economical form, since the field $\tilde{\varepsilon}_P(\omega)$ appears both in its numerator in $|\Psi_P(t)\rangle$ and denominator. To simplify it, we need to work a few more steps to cancel these field factors. Computing the Fourier transform of the polarization,

$$\begin{aligned}
\mathbf{e}_P^*\cdot\mathbf{P}_{\mathbf{k}_P}(\omega) &= \int_{-\infty}^{\infty} \mathrm{d}t\ \mathrm{e}^{\mathrm{i}\omega t}\langle\Psi_0(t)|\boldsymbol{\mu}\cdot\mathbf{e}_P^*|\Psi_P(t)\rangle \\
&= i\left[\int_0^{\infty} \mathrm{d}t\ \underbrace{\mathrm{e}^{\mathrm{i}(\omega+E_0)t}\langle\Psi_0|(\boldsymbol{\mu}\cdot\mathbf{e}_P^*)\mathrm{e}^{-\mathrm{i}H_0 t}(\boldsymbol{\mu}\cdot\mathbf{e}_P)|\Psi_0\rangle}_{(\cdot)}\right]\mathrm{e}^{\mathrm{i}\omega t_P}\tilde{\varepsilon}_P(\omega), \qquad \text{(D.7)}
\end{aligned}$$

where we have used the definition of $|\Psi_P(t)\rangle$ in equation (4.2). We also used equation (D.3*a*) to express the resulting time-domain convolution in terms of a product of functions in frequency domain. We note that $\Re\int_0^\infty dt(\cdot) = \frac{1}{2}\int_{-\infty}^{\infty} dt(\cdot)$. Substituting this result in equation (D.2), we find,

$$A_{\mathbf{k}_P}(\omega) = \frac{2\pi\omega}{c}\int_{-\infty}^{\infty} dt\, e^{i(\omega+E_0)t}\langle\Psi_0|(\boldsymbol{\mu}\cdot\mathbf{e}_P^*)e^{-iH_0t}(\boldsymbol{\mu}\cdot\mathbf{e}_P)|\Psi_0\rangle. \tag{D.8}$$

In the case that the pulse of light is ideally broadband, $\varepsilon_P(t-t_P)\propto\delta(t)$, equation (D.8) can be interpreted in the same light as equation (D.5); the wavepackets created by this field are $|\Psi_P\rangle\propto(\boldsymbol{\mu}\cdot\mathbf{e}_P)|\Psi_0\rangle$ and $|\Psi_P(t)\rangle\propto e^{-iH_0t}(\boldsymbol{\mu}\cdot\mathbf{e}_P)|\Psi_0\rangle$. Recall that the linear absorption spectrum $A_{\mathbf{k}_P}(\omega)$ is a ratio of the number of photons absorbed to the number of photons impinging on the material. By constructing this ratio, one treats every possible transition energy on the same footing, which is an equivalent way to say that a $\delta(t)$ pulse contains every frequency with the same intensity.

3. The Fourier transform in equation (D.8) can be formally evaluated by inserting a complete set of states $\mathbb{I} = \sum_m |m\rangle\langle m|$, which constitutes the vibronic eigenbasis of H_0. This yields,

$$A_{\mathbf{k}_P}(\omega) = \sum_m \frac{4\pi^2\omega}{c}|\langle m|\boldsymbol{\mu}\cdot\mathbf{e}_P|\Psi_0\rangle|^2\delta(E_m - E_0 - \omega). \tag{D.9}$$

This expression – which is essentially Fermi's golden rule – indicates that the probability that a photon of energy ω is absorbed depends on the transition amplitude $\langle m|\boldsymbol{\mu}\cdot\mathbf{e}_P|\Psi_0\rangle$ between the initial $|\Psi_0\rangle$ and the final state $|m\rangle$, provided that the transition energy $E_m - E_0$ matches the photon energy ω. The energy absorbed is proportional to this probability times the energy ω of the photon.

Although formally identical, equations (D.8) and (D.9) have different uses in practice. Equation (D.9) is a useful construct for formal arguments and for high-resolution spectroscopy of small systems, where a few eigenstates of H_0 explain the entire physics. On the other hand, complex molecular systems typically display highly congested spectra. However, if we wish to obtain an approximate $A_{\mathbf{k}_P}(\omega)$ with resolution $\Delta\omega$, we can replace the integral in equation (D.8) by a finite one, $\int_{-\infty}^{\infty}\to\int_{-T}^{T}$, where $T\sim\frac{1}{\Delta\omega}$. Hence, a short time computational simulation of duration T may yield insights on the coarse structure of $A_{\mathbf{k}_P}(\omega)$, bypassing a tedious and daunting calculation of eigenvalues and eigenfunctions [1].

Example 13. Frequency-resolved *PP′* spectroscopy

Equation (D.2) shows that an arbitrary frequency-resolved signal $A_{\mathbf{k}_s}(\omega)$ can be calculated from the Fourier transform of the corresponding time-dependent polarization $\mathbf{e}_{LO}^*\cdot\mathbf{P}_{\mathbf{k}_s}(t)$. Whether a simpler time-dependent expression for the same quantity exists in the form of a wavepacket autocorrelation function such as in the linear absorption case (equation (D.8)) is a different issue. In this example, we shall gain some intuition about frequency-resolved PP' spectroscopy by solving a simple analytical model and also deriving an associated wavepacket autocorrelation expression which is valid under certain limits, described below.

In analogy to the frequency-integrated PP' signal (see equation (4.18)), the corresponding frequency-resolved spectrum $A_{PP'}(\omega)$ is given by the differential signal,

$$A_{PP'} = A_{\mathbf{k}_{P'}}(\text{with } P) - A_{\mathbf{k}_{P'}}(\text{without } P), \tag{D.10}$$

so the relevant polarization to compute in equation (D.2) is the third order $\mathbf{e}^*_{P'} \cdot \tilde{\mathbf{P}}^{(3)}_{\mathbf{k}_{P'}}(\omega) = \int_{-\infty}^{\infty} e^{i\omega t}\mathbf{e}^*_{P'} \cdot \mathbf{P}^{(3)}_{\mathbf{k}_{P'}}(t)$, where $\mathbf{P}^{(3)}_{\mathbf{k}_{P'}}(t)$ is given by equations (3.33) or (4.19),

$$\begin{aligned} P^{(3)}_{\mathbf{k}_{P'}}(t) = & \underbrace{\langle \Psi_{+P}(t') | \boldsymbol{\mu} | \Psi_{+P+P'}(t') \rangle}_{\text{ESA}} + \underbrace{\langle \Psi_{+P-P'}(t') | \boldsymbol{\mu} | \Psi_{+P}(t') \rangle}_{\text{SE}} \\ & + \underbrace{\langle \Psi_{+P-P}(t') | \boldsymbol{\mu} | \Psi_{+P'}(t) \rangle + \langle \Psi_0(t) | \boldsymbol{\mu} | \Psi_{+P-P+P'}(t') \rangle}_{\text{GSB}}. \end{aligned} \tag{D.11}$$

This expression is valid for an arbitrary coupled dimer with vibrations given by equation (2.1), and in fact for any other molecular system. We have also noted the physical meaning of each of the terms in the sum. As in section 4.2, $\phi_s = \phi_{\text{LO}} = \phi_{P'}$, and the phases in equation (D.2) cancel out (the PP' signal does not depend on the overall phases of each pulse), so we have,

$$A_{PP'}(\omega) = \frac{4\pi\omega}{|\tilde{\varepsilon}_{P'}(\omega)|^2 c} \Im e^{-i\omega t_{\text{LO}}} \tilde{\varepsilon}^*_{P'}(\omega) \mathbf{e}^*_{P'} \cdot \tilde{\mathbf{P}}^{(3)}_{\mathbf{k}_{P'}}(\omega) \tag{D.12a}$$

$$= A_{\text{ESA}}(\omega) + A_{\text{SE}}(\omega) + A_{\text{GSB}}(\omega). \tag{D.12b}$$

where $A_{\text{ESA}}(\omega)$, $A_{\text{SE}}(\omega)$ and $A_{\text{GSB}}(\omega)$ follow from the application of equation (D.12a) to each of the terms in equation (D.11)).

1. Find explicit expressions for each of the terms in equation (D.12b) for the vibrationless coupled dimer, which is the four-level electronic system with no coupling to vibrations that we studied in section 3.2. Luckily, we have already done half of the work by computing $\mathbf{P}^{(3)}_{\mathbf{k}_{P'}}(t)$ in equations (3.34a)–(3.34d)! These equations assume the RWA and $T \gg \sigma_i$ (both of which we shall keep), but also $t \gg t_{P'}$, which can be restrictive. In order to extend their validity, let us assume that the probe pulse P' is short compared to the dynamics in each excitation manifold, but long compared to the inverse of the gap between excitation manifolds, so we may continue to use of the RWA. With this assumption, the polarization $\mathbf{P}^{(3)}_{\mathbf{k}_{P'}}(t)$ is correctly given by equations (3.34a)–(3.34d) if $t > t_{P'}$ (rather than $t \gg t_{P'}$) and is zero otherwise, indicating that the polarization exists immediately and only after the probe pulse P' acts on the material. Interpret the resulting expressions.
2. In analogy to equation (D.8), find wavepacket autocorrelation expressions for each of the terms in equation (D.12b) for a general coupled dimer *with* vibrations. For this to be possible, assume again that $T \gg \sigma_i$ and that P' is short in the same sense as indicated in part 1. Interpret the results.

Solution

1. Let us explicitly show how to compute $A_{\text{ESA}}(\omega)$ for the ideal (i.e., vibrationless) coupled dimer. Since the pulse is short compared to the dynamics in each

excitation manifold, we can take $\varepsilon_{P'}(t - t_{P'}) \approx \eta\delta(t - t_{P'})$ for the purpose of evaluation of equation (D.2). Then assuming the RWA, $\langle\Psi_{+P}(t)|\boldsymbol{\mu}|\Psi_{+P+P'}(t)\rangle$ is correctly given by equation (3.34*a*) for $t > t_{P'}$ and is zero otherwise. Therefore, we can use the dummy variable $\tilde{t} = t - t_{P'}$ to write equation (D.12*a*) as

$$A_{\text{ESA}}(\omega) = \frac{4\pi\omega}{|\tilde{\varepsilon}_{P'}(\omega)|^2 c} \Im e^{-i\omega t_{\text{LO}}} \tilde{\varepsilon}^*_{P'}(\omega) \mathbf{e}^*_{P'} \cdot \int_{-\infty}^{\infty} dt\, e^{i\omega t} \langle\Psi_{+P}(t')|\boldsymbol{\mu}|\Psi_{+P+P'}(t')\rangle$$

$$= \frac{4\pi\omega}{|\eta|^2 c} \Im\left\{ e^{-i\omega t_P} \tilde{\varepsilon}^*_{P'}(\omega) \mathbf{e}^*_{P'} \cdot (i) \sum_{p,q=\alpha,\beta} \boldsymbol{\mu}_{pf} \Omega^{P'}_{fq} \Omega^{P}_{qg} \Omega^{\overline{P}}_{gp} e^{i\omega t_{P'}} e^{-i\omega_{qp}T} \right.$$

$$\left. \times \int_0^{\infty} d\tilde{t}\, e^{i\omega\tilde{t}} e^{-i\omega_{fp}\tilde{t}} \right\}. \tag{D.13}$$

To simplify this expression, note that $\Omega^{\overline{P'}}_{pf} = \tilde{\varepsilon}^*_{P'}(\omega)\mathbf{e}^*_{P'}\boldsymbol{\mu}_{pf}$ (equation (3.22)), and regularize the integral by multiplying the integrand by a weak damping factor $e^{-\epsilon\tilde{t}}$ for $\epsilon = 0^+$,

$$A_{\text{ESA}}(\omega) = \frac{4\pi\omega}{|\eta|^2 c} \Re \sum_{p,q=\alpha,\beta} \frac{\Omega^{P'}_{fq} \Omega^{P}_{qg} \Omega^{\overline{P}}_{gp} \Omega^{\overline{P'}}_{pf} e^{-i\omega_{qp}T}}{i(\omega_{fp} - \omega - i\epsilon)}. \tag{D.14}$$

$A_{\text{ESA}}(\omega)$ consists of two resonances centered at $\omega = \omega_{f\alpha} = \omega_{\beta g}$ and $\omega = \omega_{f\beta} = \omega_{\alpha g}$, which oscillate as a function of waiting time T due to coherences $e^{-i\omega_{\alpha\beta}T}$. The processes involving populations (i.e., $p = q$) show up as typical absorptive Lorentzian lineshapes, whereas coherences (i.e., $p \neq q$) involve both absorptive and dispersive features.

Proceeding analogously with equations (3.34*b*)–(3.34*d*),

$$A_{\text{SE}}(\omega) = -\frac{4\pi\omega}{|\eta|^2 c} \Re \sum_{p,q=\alpha,\beta} \frac{\Omega^{\overline{P'}}_{gq} \Omega^{P}_{qg} \Omega^{\overline{P}}_{gp} \Omega^{P'}_{pg} e^{-i\omega_{qp}T}}{i(\omega_{pg} - \omega - i\epsilon)}, \tag{D.15}$$

$$A_{\text{GSB}}(\omega) = -\frac{4\pi\omega}{|\eta|^2 c} \Re \sum_{p,q=\alpha,\beta} \frac{\Omega^{\overline{P'}}_{gq} \Omega^{P'}_{qg} \Omega^{\overline{P}}_{gp} \Omega^{P}_{pg}}{i(\omega_{qg} - \omega - i\epsilon)}. \tag{D.16}$$

The total signal is depicted in figure D.1. See appendix E for a signal containing similar information to $A_{PP'}(\omega)$ in the photon-echo context. Note the negative signs in front of these last expressions compared to their ESA counterpart, just like the corresponding frequency-integrated signals. Importantly, due to the coherences, these signs do not preclude negative features in A_{ESA} or positive ones in $A_{\text{SE}}(\omega)$! This contrasts strongly with the frequency-integrated signals S_{ESA} and S_{SE}, which are always positive and negative contributions, respectively

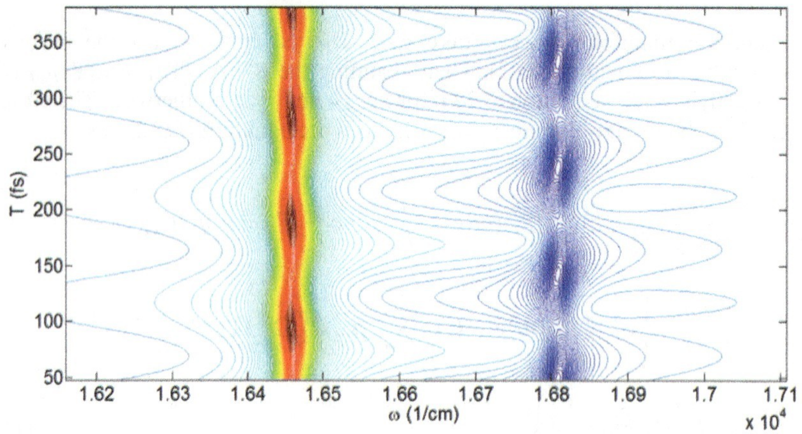

Figure D.1. Frequency resolved pump–probe spectrum $A_{PP'}(\omega)$ for an ideal coupled dimer as a function of waiting time T. The spectrum consists of two resonances centered at $\omega_{\alpha g} = \omega_{f\beta} = 16\,458$ cm^{-1} and $\omega_{\beta g} = \omega_{\alpha g} = 16\,808$ cm^{-1}, respectively (the sites have identical energies and the coupling between them is $J = 175$ cm^{-1}). Coherences between $|\alpha\rangle$ and $|\beta\rangle$ modulate the spectra as a function of waiting time T with a period of oscillations of 95 fs. The results have been isotropically averaged for a cross-polarized configuration **xxzz**.

(see example 6). The reason is that in any process and at every photon frequency ω, energy can be given to or taken away from the field, but the total number of photons irrespective of their color is always given to the field in a SE process and taken away from it in an ESA process.

As a consistency check, just as in example 12, the frequency integrated signal can be recovered by dividing the frequency-resolved spectrum by the photon frequency and integrating over all frequencies, for instance, $\int_{-\infty}^{\infty} \mathrm{d}\omega \frac{A_{\mathrm{ESA}}}{\omega} \propto \Re \sum_{p,q=\alpha,\beta} \Omega_{gq}^{\overline{P}'} \Omega_{qg}^{P} \Omega_{gp}^{\overline{P}} \Omega_{pg}^{P'} \mathrm{e}^{-\mathrm{i}\omega_{qp}T} \propto S_{\mathrm{ESA}}$ (see equation (4.44a)). Here, we have used,

$$\frac{1}{\mathrm{i}(\omega_{ij} - \omega - \mathrm{i}\epsilon)} = \frac{-\mathrm{i}(\omega_{ij} - \omega)}{(\omega_{ij} - \omega)^2 + \epsilon^2} + \frac{\epsilon}{(\omega_{ij} - \omega)^2 + \epsilon^2}, \tag{D.17}$$

where the first and second terms on the right hand side are the dispersive and absorptive parts of the lineshape. The dispersive part is odd with respect to ω so it does not contribute to the signal. The absorptive part is a Lorentzian, which tends to $\pi\delta(\omega_{ij} - \omega)$ as $\epsilon \to 0^+$.

2. Let us now generalize the results from part 1 to a coupled dimer with vibrations, still using the broadband limit for P'. As in exercise 12, equation (D.12a) can be reexpressed in the time domain using the convolution theorem (equation (D.3)),

$$\begin{aligned} A_{PP'}(\omega) &= \frac{4\pi\omega}{|\tilde{\varepsilon}_{P'}(\omega)|^2 c} \Im\{\mathrm{e}^{-\mathrm{i}\omega t_{\mathrm{LO}}} \tilde{\varepsilon}_{P'}^*(\omega) \mathbf{e}_{P'}^* \cdot \tilde{\mathbf{P}}_{\mathbf{k}_{P'}}^{(3)}(\omega)\} \\ &= \frac{4\pi\omega}{|\tilde{\varepsilon}_{P'}(\omega)|^2 c} \Im \int_{-\infty}^{\infty} \mathrm{d}t\, \mathrm{e}^{\mathrm{i}\omega t} \int_{-\infty}^{\infty} \mathrm{d}t' \varepsilon_{P'}^*(t' - t - t_{P'}) \cdot \mathbf{e}_{P'}^* \tilde{\mathbf{P}}_{\mathbf{k}_{P'}}^{(3)}(t'). \end{aligned} \tag{D.18}$$

The integrals corresponding to $A_{\mathrm{ESA}}(\omega)$ are,

$$
\begin{aligned}
&\int_{-\infty}^{\infty} \mathrm{d}t\, \mathrm{e}^{\mathrm{i}\omega t} \int_{-\infty}^{\infty} \mathrm{d}t' \varepsilon^*_{P'}(t' - t - t_{P'}) \cdot \mathbf{e}^*_{P'} \langle \Psi_{+P}(t') | \boldsymbol{\mu} | \Psi_{+P+P'}(t') \rangle \\
&\quad = |\eta|^2 \int_{-\infty}^{\infty} \mathrm{d}t\, \mathrm{e}^{\mathrm{i}\omega t} \int_{-\infty}^{\infty} \mathrm{d}t' \int_{-\infty}^{t'} \mathrm{d}t'' \delta(t' - t - t_{P'}) \delta(t'' - t_{P'}) \\
&\qquad \times \langle \Psi_{+P}(t') | \boldsymbol{\mu} \cdot \mathbf{e}^*_{P'} \mathrm{e}^{-\mathrm{i}H_0(t'-t'')} \mathbb{P}_{\mathrm{DEM}}(\mathrm{i}) \boldsymbol{\mu} \cdot \mathbf{e}_{P'} | \Psi_{+P}(t'') \rangle \\
&\quad = \mathrm{i}|\eta|^2 \int_{-\infty}^{\infty} \mathrm{d}t\, \mathrm{e}^{\mathrm{i}\omega t} \int_{-\infty}^{t+t_{P'}} \mathrm{d}t'' \delta(t'' - t_{P'}) \\
&\qquad \times \langle \Psi_{+P}(t + t_{P'}) | \boldsymbol{\mu} \cdot \mathbf{e}^*_{P'} \mathrm{e}^{-\mathrm{i}H_0(t+t_{P'}-t'')} \mathbb{P}_{\mathrm{DEM}} \boldsymbol{\mu} \cdot \mathbf{e}_{P'} | \Psi_{+P}(t'') \rangle \\
&\quad = \mathrm{i}|\eta|^2 \int_0^{\infty} \mathrm{d}t\, \mathrm{e}^{\mathrm{i}\omega t} \langle \Psi_{+P}(t_{P'}) | \mathrm{e}^{\mathrm{i}H_0 t} \boldsymbol{\mu} \cdot \mathbf{e}^*_{P'} \mathrm{e}^{-\mathrm{i}H_0 t} \mathbb{P}_{\mathrm{DEM}} \boldsymbol{\mu} \cdot \mathbf{e}_{P'} | \Psi_{+P}(t_{P'}) \rangle
\end{aligned}
$$

From the first to the second line, we have rewritten $|\Psi_{+P+P'}(t')\rangle$ as the result of the $\varepsilon_{P'}$ perturbation on $|\Psi_{+P}(t'')\rangle$, keeping track of the factor of i that comes with it. The RWA guarantees that $|\Psi_{+P+P'}(t')\rangle$ is in the DEM even in the limit of $\varepsilon_{P'}$ becoming a δ-function. From the second to the third line we have evaluated the t' integral, and finally, from the third to the fourth line we have evaluated the t'' integral, taking care of the fact that it vanishes for $t < 0$, so we end up with only half of the t integral, $\int_{-\infty}^{\infty} \mathrm{d}t \to \int_0^{\infty} \mathrm{d}t$. Similar manipulations can be carried out for the $A_{\mathrm{SE}}(\omega)$ and $A_{\mathrm{GSB}}(\omega)$ terms. The final results are,

$$
A_{\mathrm{ESA}}(\omega) = \frac{4\pi\omega}{c} \Re \int_0^{\infty} \mathrm{d}t\, \mathrm{e}^{\mathrm{i}\omega t} \langle \Psi_{+P}(t_{P'}) | \mathrm{e}^{\mathrm{i}H_0 t} \boldsymbol{\mu} \cdot \mathbf{e}^*_{P'} \mathrm{e}^{-\mathrm{i}H_0 t} \mathbb{P}_{\mathrm{DEM}} \boldsymbol{\mu} \cdot \mathbf{e}_{P'} | \Psi_{+P}(t_{P'}) \rangle, \tag{D.19}
$$

$$
A_{\mathrm{SE}}(\omega) = -\frac{4\pi\omega}{c} \Re \int_0^{\infty} \mathrm{d}t\, \mathrm{e}^{\mathrm{i}\omega t} \langle \Psi_{+P}(t_{P'}) | \boldsymbol{\mu} \cdot \mathbf{e}^*_{P'} \mathrm{e}^{\mathrm{i}H_0 t} \mathbb{P}_{\mathrm{GSM}} \boldsymbol{\mu} \cdot \mathbf{e}_{P'} \mathrm{e}^{-\mathrm{i}H_0 t} | \Psi_{+P}(t_{P'}) \rangle, \tag{D.20}
$$

$$
\begin{aligned}
A_{\mathrm{GSB}}(\omega) = \frac{4\pi\omega}{c} \Re \Bigg\{ & \int_0^{\infty} \mathrm{d}t\, \mathrm{e}^{\mathrm{i}\omega t} \langle \Psi_0(t_{P'}) | \mathrm{e}^{\mathrm{i}H_0 t} \boldsymbol{\mu} \cdot \mathbf{e}^*_{P'} \mathrm{e}^{-\mathrm{i}H_0 t} \mathbb{P}_{\mathrm{SEM}} \boldsymbol{\mu} \cdot \mathbf{e}_{P'} | \Psi_{+P-P}(t_{P'}) \rangle \\
&+ \int_0^{\infty} \mathrm{d}t\, \mathrm{e}^{\mathrm{i}\omega t} \langle \Psi_{+P-P}(t_{P'}) | \mathrm{e}^{\mathrm{i}H_0 t} \boldsymbol{\mu} \cdot \mathbf{e}^*_{P'} \mathrm{e}^{-\mathrm{i}H_0 t} \mathbb{P}_{\mathrm{SEM}} \boldsymbol{\mu} \cdot \mathbf{e}_{P'} | \Psi_0(t_{P'}) \rangle \Bigg\}.
\end{aligned} \tag{D.21}
$$

$A_{\mathrm{ESA}}(\omega)$ can be interpreted as follows. In general, $|\Psi_{+P}\rangle$ is a nonstationary wavepacket that spends approximately[1] $T = t_{P'} - t_P$ in the SEM. At time $t_{P'}$, the probe arrives and a 'frozen' copy of the wavepacket $|\Psi_{+P}(t_{P'})\rangle$ is taken up to the DEM and allowed to

[1] The evolution time is only approximate because ε_P may have a finite width in time.

evolve under H_{DEM}. The leftover copy in the SEM continues evolving under H_{SEM}. $A_{\mathrm{ESA}}(\omega)$ is the Fourier transform of the overlap between these two wavepackets. To make an analogy with linear absorption, in the case that $|\Psi_{+P}(t_P)\rangle$ is a stationary state, $\langle\Psi_{+P}(t_{P'})|e^{iH_0 t} = \langle\Psi_{+P}(t_P)|e^{iE_P t}$, and $A_{\mathrm{ESA}}(\omega)$ acquires the same form as equation (D.8), being just the linear absorption spectrum of a SEM state being promoted to the DEM. However, $\langle\Psi_{+P}(t_{P'})|$ is not a stationary state in general, and will be in different positions at different t_P or T values.

The interpretation of $A_{\mathrm{SE}}(\omega)$ follows analogously to $A_{\mathrm{ESA}}(\omega)$. Here, the minus sign is expected compared to its ESA counterpart. The initial state in the SEM has more energy than the final state in the GSM, so the signs of the propagators are switched compared to the $A_{\mathrm{ESA}}(\omega)$ case. Finally, $A_{\mathrm{GSB}}(\omega)$ can also be understood in similar fashion.[2]

Equations (D.19)–(D.21) are very amenable to numerical simulation and can be computed by adapting the example MATLAB® code provided with this book. When using them, we just need to keep in mind their limitations coming from the assumptions we considered to derive them. As a check, the reader might want to confirm that when applied to the vibrationless coupled dimer, they give back the same expressions as equations (D.14)–(D.16). These results are closely related to expressions which, as far as we are aware, were first derived by Pollard and coworkers [2, 3], who refer to frequency-resolved PP' spectroscopy as 'dynamic absorption.'

Bibliography

[1] Heller E J 1981 The semiclassical way to molecular spectroscopy *Acc. Chem. Res.* **14** 368–75

[2] Pollard W T, Lee S Y and Mathies R A 1990 Wave packet theory of dynamic absorption spectra in femtosecond pump–probe experiments *J. Chem. Phys.* **92** 4012–29

[3] Pollard W T and Mathies R A 1992 Analysis of femtosecond dynamic absorption spectra of nonstationary states *Ann. Rev. Phys. Chem.* **43** 497–523

[2] Recall that GSB corresponds to the change in absorption of P' from the GSM to the SEM due to the reduced population in the GSM depleted after P (see exercise 5). $|\Psi_{+P-P}(t_{P'})\rangle$ is a hole wavefunction (see example 6), coming with a negative sign with respect to the zeroth order wavefunction $|\Psi_0(t_{P'})\rangle$ due to the two P perturbations ($(i)(i) = -1$). Though we generally expect the SE and GSB terms to have the same sign, $|\Psi_{+P-P}\rangle$ being a negative contribution explains why $A_{\mathrm{GSB}}(\omega)$ does not have an explicit minus sign as $A_{\mathrm{SE}}(\omega)$ does. This is entirely analogous to the signs in the wavepacket overlap expressions for S_{GSB} and S_{SE} in equations (4.28*b*) and (4.29*b*), respectively.

IOP Publishing

Ultrafast Spectroscopy
Quantum Information and Wavepackets
Joel Yuen-Zhou *et al*

Appendix E

Two-dimensional spectroscopy

We now develop a simple example of photon-echo (PE) spectroscopy, yet another example of four-wave mixing. The setup is the same as for TG spectroscopy (see example 10), where four pulses satisfying the condition $\mathbf{k}_4 = \mathbf{k}_{\mathrm{PE}} \equiv -\mathbf{k}_1 + \mathbf{k}_2 + \mathbf{k}_3$ interact with the sample, and we are interested in the third order signal in the $\mathbf{k}_{\mathrm{PE}}$ $(= \mathbf{k}_{\mathrm{TG}})$ direction. However, as opposed to TG, we consider the case where $t_1 < t_2 < t_3 < t_4$, in which the different time intervals are labeled by the names *coherence* time $\tau \equiv t_2 - t_1$, *waiting* time $T = t_3 - t_2$, and *echo* time $\bar{t} = t_4 - t_3$ [5, 14]. In PP' and TG spectroscopy, $\tau = \bar{t} = 0$ and the only relevant time interval is T. In PE spectroscopy, we will take advantage of all three intervals, and assume that $\{\tau, T, \bar{t}\} \gg \sigma_i$, that is, the pulses are well separated. This allows us to obtain different expressions from those of the TG signal (equations (5.36)–(5.40)). See figure E.1.

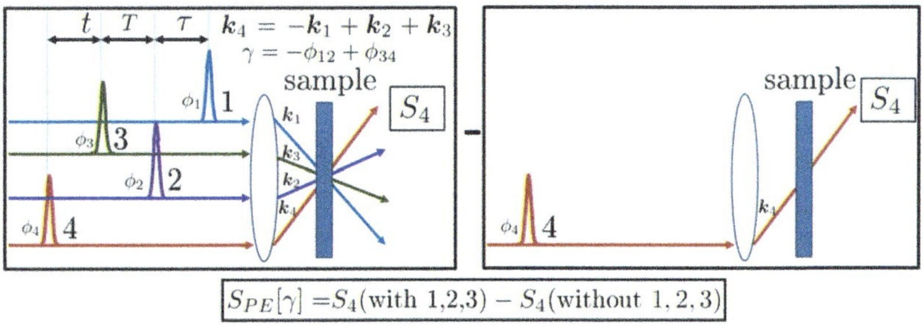

Figure E.1. Photon-echo (PE) spectroscopy. Four noncollinear pulses with wavevector constraint $\mathbf{k}_{\mathrm{PE}} = \mathbf{k}_4 = -\mathbf{k}_1 + \mathbf{k}_2 + \mathbf{k}_3$ interact with the sample. Only the differential absorption of pulse 4 is monitored. The pulses are separated in time by intervals $\tau = t_2 - t_1$, $T = t_3 - t_2$ and $\bar{t} = t_4 - t_3$, called the coherence, waiting and echo times, respectively. When $\tau = \bar{t} = 0$, the PE setup reduces to a TG experiment (see figure 5.7).

Let us generalize the expressions for $S_{\mathrm{TG}}[\gamma]$ (equations (5.28) and (5.30)) to the PE context,

$$\begin{aligned} S_{\mathrm{PE}}[\gamma](\tau,T,\bar{t}) &= S_4 \text{ (with } 1,2,3) - S_4 \text{ (without } 1,2,3) \\ &= 2\Im \int_{-\infty}^{\infty} \mathrm{d}t' \varepsilon_4^*(t'-t_4)\mathbf{e}_4^* \cdot \mathbf{P}^{(3)}_{\mathbf{k}_{\mathrm{PE}}}(t')\mathrm{e}^{\mathrm{i}\gamma} \\ &= S_{\mathrm{ESA}}[\gamma](\tau,T,\bar{t}) + S_{\mathrm{SE}}[\gamma](\tau,T,\bar{t}) + S_{\mathrm{GSB}}[\gamma](\tau,T,\bar{t}), \end{aligned} \tag{E.1}$$

where, so far, the only thing we have done is to change the subscript from TG to PE. The expression for the third-order polarization $\mathbf{P}^{(3)}_{\mathbf{k}_{\mathrm{PE}}}(t)$ is equivalent to that of $\mathbf{P}^{(3)}_{\mathbf{k}_{\mathrm{TG}}}(t)$ in equation (5.34). We begin by adapting equation (5.35) to obtain an expression for the ESA component of the PE signal,

$$\begin{aligned} S_{\mathrm{ESA}}[\gamma] &= 2\Im \mathrm{e}^{\mathrm{i}\gamma} \int_{-\infty}^{\infty} \mathrm{d}t' \varepsilon_4^*(t'-t_4)\langle \Psi_{+1}(t')|\boldsymbol{\mu}\cdot\mathbf{e}_4^*|\Psi_{+2+3}(t')\rangle \\ &= 2\Im \left\{ \underbrace{\mathrm{e}^{\mathrm{i}\gamma}\mathrm{i}\int_{-\infty}^{\infty} \mathrm{d}t' \langle \Psi_1(t')|(-\mathrm{i})\varepsilon_4^*(t'-t_4)\boldsymbol{\mu}\cdot\mathbf{e}_4^*\mathrm{e}^{\mathrm{i}H_0(t-t')}}_{\approx\langle\Psi_{14}(t)|} \right. \\ &\quad \left. \times \underbrace{\mathrm{e}^{-\mathrm{i}H_0(t-t')}\mathbb{P}_{\mathrm{DEM}}|\Psi_{23}(t')\rangle}_{=\mathbb{P}_{\mathrm{DEM}}|\Psi_{23}(t)\rangle} \right\} \\ &\approx 2\Re \mathrm{e}^{\mathrm{i}\gamma}\langle\Psi_{14}|\mathbb{P}_{\mathrm{DEM}}|\Psi_{23}\rangle, \end{aligned} \tag{E.2}$$

where in the second line we have inserted the identity in the form $\mathbb{I} = \mathrm{e}^{-\mathrm{i}H_0(t-t')}\mathrm{e}^{\mathrm{i}H_0(t-t')}$ and chosen t such that $t - t_4 \gg \sigma$. Crucially, we have approximated $|\Psi_{14}(t)\rangle$ by taking the integral $\int_{-\infty}^{t} \mathrm{d}t' \approx \int_{-\infty}^{\infty} \mathrm{d}t'$, which holds as long as pulse 4 is well separated from the other pulses, $(t_4 - t_i)/\sigma \gg 1$ for $i = 1, 2, 3$. This approximation was inapplicable in the context of the TG setup, where pulses 3 and 4 arrive at the same time. The reason for this discrepancy can be understood in simple physical terms. The energy lost by pulse 4 in the TG setup depends on how pulses 3 and 4 transfer amplitude simultaneously from the SEM to the DEM (see equation (5.46)). In contrast, in the PE setup, the wavepackets $|\Psi_{14}\rangle$ and $|\Psi_{23}\rangle$ are born 'independently' at different instants t_4 and t_3, respectively, and the energy lost by pulse 4 depends on the interference between the two resulting wavepackets.

In a similar fashion, one finds

$$S_{\mathrm{SE}}[\gamma](\tau, T, \bar{t}) \equiv -2\Re \mathrm{e}^{\mathrm{i}\gamma}\langle\Psi_{13}|\mathbb{P}_{\mathrm{GSM}}|\Psi_{24}\rangle, \tag{E.3}$$

$$S_{\mathrm{GSB}}[\gamma](\tau, T, \bar{t}) = 2\Re \mathrm{e}^{\mathrm{i}\gamma}[\langle\Phi_{124}|\Psi_3\rangle]. \tag{E.4}$$

So far, equations (E.2)–(E.4) are very general. Similar expressions have been reported in the context of nonlinear wavepacket interferometry [1, 2, 6–8, 11, 12]). Let us now calculate the different wavepackets needed to evaluate them for the particular case of the vibrationless coupled dimer, which is a four-level electronic system (section 3.2). The results from equations (4.41)–(4.43) can be readily adapted without much more calculation,

$$|\Psi_m\rangle = \mathrm{i}\sum_{q=\alpha,\beta} \mathrm{e}^{\mathrm{i}\omega_q t_m}|q\rangle\Omega^m_{qg}\mathrm{e}^{-\mathrm{i}\omega_g t_m}, \tag{E.5}$$

$$\begin{aligned}|\Psi_{mn}\rangle = &-\mathrm{e}^{\mathrm{i}\omega_f t_n}|f\rangle\sum_{q=\alpha,\beta}\Omega^n_{fq}\mathrm{e}^{-\mathrm{i}\omega_q(t_n-t_m)}\Omega^m_{qg}\mathrm{e}^{-\mathrm{i}\omega_g t_m}\\ &-\mathrm{e}^{\mathrm{i}\omega_g t_n}|g\rangle\sum_{q=\alpha,\beta}\Omega^{\bar{n}}_{gq}\mathrm{e}^{-\mathrm{i}\omega_q(t_n-t_m)}\Omega^m_{qg}\mathrm{e}^{-\mathrm{i}\omega_g t_m}\end{aligned} \tag{E.6}$$

$$|\Phi_{mnr}\rangle = -\mathrm{i}\sum_{p,q=\alpha,\beta}|p\rangle\mathrm{e}^{\mathrm{i}\omega_p t_r}\Omega^r_{pg}\mathrm{e}^{-\mathrm{i}\omega_g(t_r-t_n)}\Omega^{\bar{n}}_{gq}\mathrm{e}^{-\mathrm{i}\omega_q(t_n-t_m)}\Omega^m_{qg}\mathrm{e}^{-\mathrm{i}\omega_g t_m} \tag{E.7}$$

where we have assumed that m, n and r are different pulses such that $t_m < t_n < t_r$ and $m, n, r \in \{1, 2, 3, 4\}$. Equation (E.7) differs by a factor of 2 from equation (4.43), and this can be understood as follows. $|\Phi_{mnr}\rangle$ indicates a wavepacket that chronologically goes from the GSM to the SEM via m, from the SEM back to the GSM via n, and from the GSM again to the SEM via r. When m and n are well separated in time, as in PE spectroscopy, the first and second pulses act in their entirety to induce each of the mentioned transitions. In the case of PP' or TG spectroscopy, the first two pulses arrive at the same time, so pulse 2 acts before pulse 1 finishes promoting its corresponding amplitude, and therefore, only half of the amplitude from the PE case is obtained.

Substituting equations (E.5)–(E.7) into equations (E.2)–(E.4),

$$S^\gamma_{\mathrm{ESA}}[\gamma](\tau,T,\bar{t}) = \sum_{p,q=\alpha,\beta}\Omega^3_{fq}\Omega^2_{qg}\Omega^{\bar{1}}_{gp}\Omega^{\bar{4}}_{pf}\mathrm{e}^{\mathrm{i}\gamma}\mathrm{e}^{-\mathrm{i}\omega_{fp}\bar{t}}\mathrm{e}^{-\mathrm{i}\omega_{qp}T}\mathrm{e}^{-\mathrm{i}\omega_{gp}\tau} + \mathrm{c.c.}, \tag{E.8a}$$

$$S^\gamma_{\mathrm{SE}}[\gamma](\tau,T,\bar{t}) = -\sum_{p,q=\alpha,\beta}\Omega^{\bar{4}}_{gq}\Omega^2_{qg}\Omega^{\bar{1}}_{gp}\Omega^3_{pg}\mathrm{e}^{\mathrm{i}\gamma}\mathrm{e}^{-\mathrm{i}\omega_{qg}\bar{t}}\mathrm{e}^{-\mathrm{i}\omega_{qp}T}\mathrm{e}^{-\mathrm{i}\omega_{gp}\tau} + \mathrm{c.c.}, \tag{E.8b}$$

$$S^\gamma_{\mathrm{GSB}}[\gamma](\tau,T,\bar{t}) = -\sum_{p,q=\alpha,\beta}\Omega^3_{qg}\Omega^{\bar{1}}_{gp}\Omega^2_{pg}\Omega^{\bar{4}}_{gq}\mathrm{e}^{\mathrm{i}\gamma}\mathrm{e}^{-\mathrm{i}\omega_{qg}\bar{t}}\mathrm{e}^{-\mathrm{i}\omega_{gp}\tau} + \mathrm{c.c.} \tag{E.8c}$$

See figure E.2 for the respective DS-FDs for equations (E.8*a*)–(E.8*c*). The coherence elements in the coherence $|g\rangle\langle p|$ oscillate at $\mathrm{e}^{-\mathrm{i}\omega_{gp}\tau}$ whereas those of the echo times $|q\rangle\langle g|$ and $|f\rangle\langle p|$ oscillate as $\mathrm{e}^{-\mathrm{i}\omega_{qg}\bar{t}}$ and $\mathrm{e}^{-\mathrm{i}\omega_{fp}\bar{t}}$, respectively. $\omega_{gp} < 0$ while ω_{qg}, $\omega_{fp} > 0$: these different signs in the energy differences earn the signal $S_{\mathrm{PE}}[\gamma](\tau, T, \bar{t})$ its adjective of *rephasing*. When $t_2 < t_1$ (not treated in this example), the signal is known as a nonrephasing PE signal. For more information, see [4, 5].

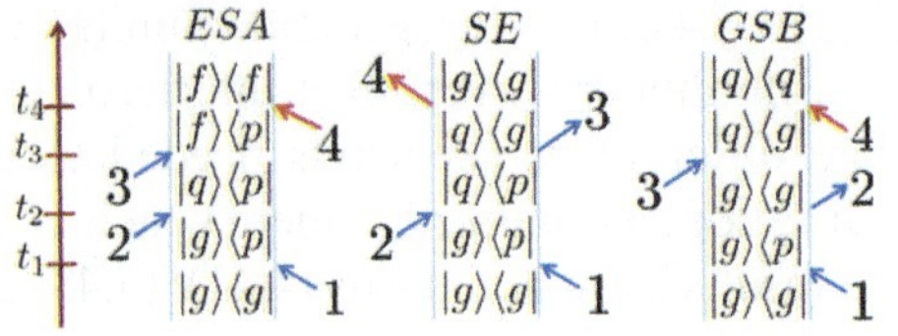

Figure E.2. DS-FDs for the ideal coupled dimer photon-echo signal S_{PE}.

We will now consider the complex-valued signal,

$$\Sigma_{\mathrm{PE}}(\tau, T, \bar{t}) \equiv \frac{1}{2}\left\{S_{\mathrm{PE}}[0](\tau, T, \bar{t}) + \mathrm{i}S_{\mathrm{PE}}\left[-\frac{\pi}{2}\right](\tau, T, \bar{t})\right\} \tag{E.9a}$$

$$= -\mathrm{i}\int_{-\infty}^{\infty} \mathrm{d}t' \varepsilon_4^*(t' - t_4)\mathbf{e}_4^* \cdot \mathbf{P}_{\mathbf{k}_{\mathrm{PE}}}^{(3)}(t'), \tag{E.9b}$$

where the second identity follows directly from equation (E.1). This signal is collected for many $\tau, T, \bar{t}$ times, and the resulting data are Fourier transformed along the coherence and echo times [5, 13],

$$S_{2D}(\omega_\tau, T, \omega_t) = \int_0^\infty \mathrm{d}\tau\, \mathrm{e}^{-\mathrm{i}\omega_\tau\tau} \int_0^\infty \mathrm{d}\bar{t}\, \mathrm{e}^{\mathrm{i}\omega_t\bar{t}} \Sigma_{\mathrm{PE}}(\tau, T, \bar{t})\mathrm{e}^{-\Gamma_\tau\tau}\mathrm{e}^{-\Gamma_t\bar{t}}$$

$$= \sum_{m,n=\alpha,\beta} l_{\tau,m}(\omega_\tau) l_{t,n}(\omega_t) S_{mn}(T). \tag{E.10}$$

where, in order to obtain sensible Fourier transforms, we have multiplied the time signal by an artificial exponentially decaying envelope $\mathrm{e}^{-\Gamma_\tau\tau}\mathrm{e}^{-\Gamma_t\bar{t}}$, where Γ_τ and Γ_t are infinitesimally small positive numbers. We would not have needed to do this artificial signal processing for cases other than the vibrationless coupled dimer, as in most cases, the coupling of system $\mathscr{S}$ to bath $\mathscr{B}$ quenches the signal as a function of time.

What we have achieved in equation (E.10) is a Two-Dimensional-Electronic Spectrum (2D-ES) along the ω_τ and ω_t axes, which consists of four peaks located at $(\omega_\tau, \omega_t) = (\omega_i, \omega_j)$, where $\omega_i, \omega_j \in \{\omega_{\alpha g}, \omega_{\beta g}\}$. Each of the peaks is associated with the product of a T-dependent amplitude $S_{qp}(T)$ [16],

$$S_{\alpha\alpha}(T) = \underbrace{-\Omega^{\bar{4}}_{g\alpha}\Omega^{2}_{\alpha g}\Omega^{\bar{1}}_{g\alpha}\Omega^{3}_{\alpha g}}_{\text{from SE}}\ \underbrace{-\Omega^{3}_{\alpha g}\Omega^{\bar{1}}_{g\alpha}\Omega^{2}_{\alpha g}\Omega^{\bar{4}}_{g\alpha}}_{\text{from GSB}}, \tag{E.11}$$

$$S_{\alpha\beta}(T) = \underbrace{\Omega^{3}_{f\alpha}\Omega^{2}_{\alpha g}\Omega^{\bar{1}}_{g\alpha}\Omega^{\bar{4}}_{\alpha f} + \Omega^{3}_{f\beta}\Omega^{2}_{\beta g}\Omega^{\bar{1}}_{g\alpha}\Omega^{\bar{4}}_{\alpha f}\mathrm{e}^{-\mathrm{i}\omega_{\beta\alpha}T}}_{\text{from ESA}}$$

$$\underbrace{-\Omega^{\bar{4}}_{g\beta}\Omega^{2}_{\beta g}\Omega^{\bar{1}}_{g\alpha}\Omega^{3}_{\alpha g}\mathrm{e}^{-\mathrm{i}\omega_{\beta\alpha}T}}_{\text{from SE}}\ \underbrace{-\Omega^{3}_{\beta g}\Omega^{\bar{1}}_{g\alpha}\Omega^{2}_{\alpha g}\Omega^{\bar{4}}_{g\beta}}_{\text{from GSB}}, \tag{E.12}$$

with $S_{\beta\beta}(T)$ and $S_{\alpha\beta}(T)$ following from the substitutions $(\alpha, \beta) \to (\beta, \alpha)$ from $S_{\alpha\alpha}(T)$ and $S_{\beta\alpha}(T)$, respectively. These amplitudes are spread out by the product of two lineshapes, each associated with a Fourier transform along τ or t:

$$l_{\tau,p}(\omega_\tau) \approx \int_0^\infty \mathrm{d}\tau\, \mathrm{e}^{-\mathrm{i}\omega_{gp}\tau}\mathrm{e}^{-\Gamma_\tau\tau}\mathrm{e}^{-\mathrm{i}\omega_\tau\tau} = \frac{1}{\mathrm{i}(\omega_\tau - \omega_{pg} - \mathrm{i}\Gamma_\tau)}, \tag{E.13}$$

$$l_{t,q}(\omega_{\bar{t}}) \approx \int_0^\infty \mathrm{d}t\, \mathrm{e}^{-\mathrm{i}\omega_{qg}\bar{t}}\mathrm{e}^{-\Gamma_{\bar{t}}\bar{t}}\mathrm{e}^{\mathrm{i}\omega_t\bar{t}} = \frac{1}{\mathrm{i}(-\omega_t + \omega_{qg} - \mathrm{i}\Gamma_{\bar{t}})}. \tag{E.14}$$

These lineshapes are approximate rather than exact because equations (E.8*a*)–(E.8*c*) are not valid for short $0 < \tau, t < \sigma_i$, yet the required integrals go from 0 to ∞. As long as the pulses are short compared to the corresponding dynamics in the coherence and echo intervals, these errors should be negligible. Notice the similarity between these expressions and the ones for frequency-resolved PP' (example 14 in appendix D).

The diagonal peaks $S_{\alpha\alpha}(T)$ and $S_{\beta\beta}(T)$ are, as in equation (E.11), T-independent signals that represent the populations $|\alpha\rangle\langle\alpha|$ and $|\beta\rangle\langle\beta|$. They also include the background signal due to GSB (associated with the δ_{pq} term). The cross-peaks $S_{\alpha\beta}(T)$ and $S_{\beta\alpha}(T)$ are oscillatory and represent the T-dependent coherent beats due to $|\alpha\rangle\langle\beta|$ and $|\beta\rangle\langle\alpha|$. The SEM dynamics can be qualitatively tracked by following the evolution of example peaks. Figure E.3 shows the corresponding 2D-ES for this model, for a particular set of typical parameters for dipoles and energy levels. Here, $T_c = \frac{2\pi}{\omega_{\alpha\beta}}$ is the period that the coherence between $|\alpha\rangle$ and $|\beta\rangle$ takes to evolve its phase by 2π. Snapshots of the spectra as a function of waiting time T are taken every half period.

The cross-peak $S_{\alpha\beta}(T)$ is sensitive to the coupling J. In the limiting case when the coupling J between the chromophore sites vanishes, we have two uncoupled two-level systems; in such a case the SEM states $|\alpha\rangle = |a\rangle$ and $|\beta\rangle = |b\rangle$ are excitations in independent sites. Regardless of the coupling, the DEM can be written as $|f\rangle = |\alpha\rangle|\beta\rangle = |a\rangle|b\rangle$. It follows that $\boldsymbol{\mu}_{f\alpha} = \boldsymbol{\mu}_{\beta g}$ and $\boldsymbol{\mu}_{f\beta} = \boldsymbol{\mu}_{\alpha g}$, and the first and fourth terms of equation (E.12) cancel exactly, and so do the second and the third. The physical content of this result is remarkable: for every SE or GSB process involving a coherence in the SEM, there is a degenerate ESA process with precisely the opposite energy and amplitude. This is the same as saying that the cross-peaks vanish. The diagonal peaks (equation (E.11)), however, do not involve ESA processes and do not vanish even in the absence of coupling J. Hence, 2D-ES is an invaluable tool to probe couplings between chromophores, and has been successfully applied to unravel the various couplings between chromophores in

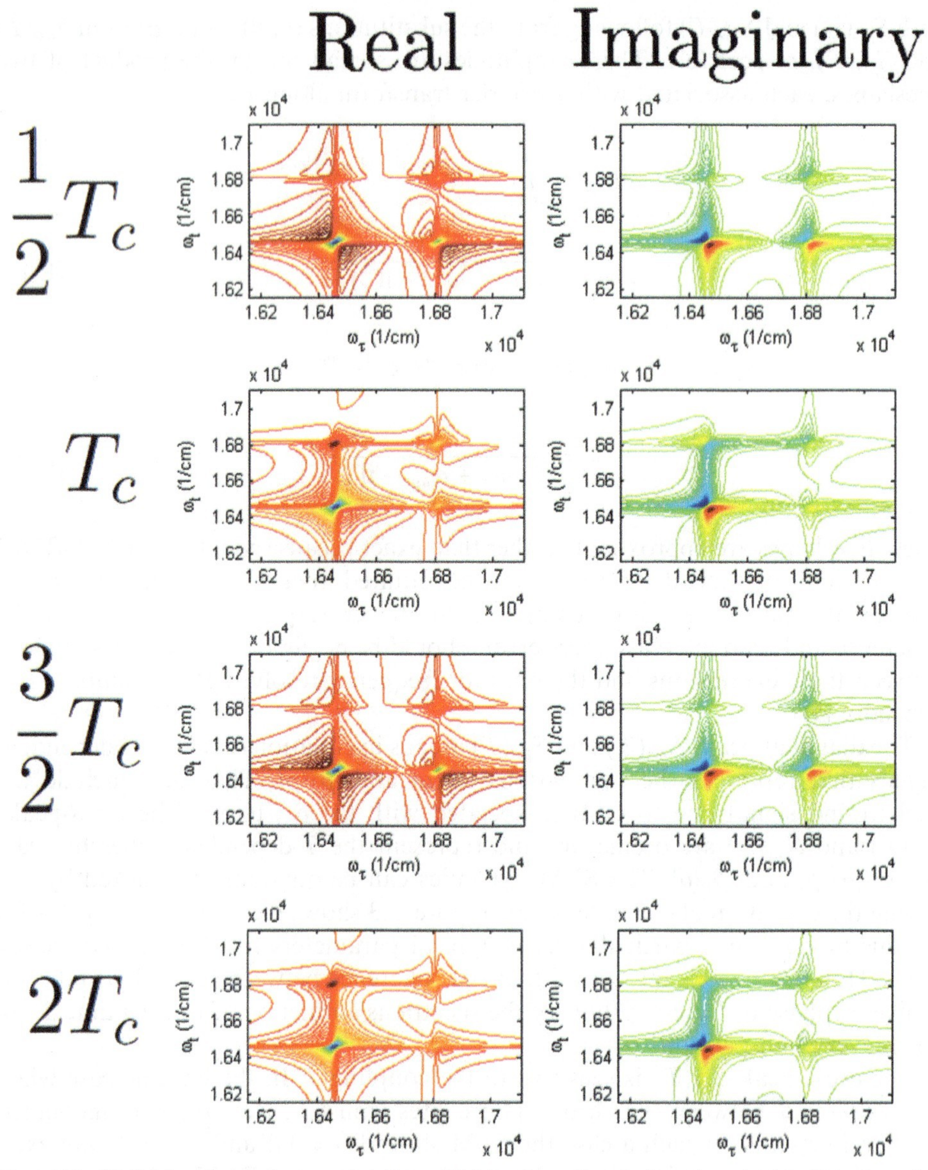

Figure E.3. 2D-ES for vibrationless coupled dimer. Waiting times T are indicated on the left. T_c is the period of one cycle of the evolution of a coherent superposition between $|\alpha\rangle$ and $|\beta\rangle$. The real and imaginary parts are shown in different columns. Cross-peaks oscillate as $e^{\mp i\omega_{\alpha\beta}T}$. The parameters here are $\omega_{\alpha g} = \omega_{f\beta} = 16\,458\,\mathrm{cm}^{-1}$, $\omega_{\beta g} = \omega_{\alpha g} = 16\,808\,\mathrm{cm}^{-1}$ and $J = 175\,\mathrm{cm}^{-1}$, as in figure D.1. The results have been isotropically averaged for a cross-polarized configuration **xxzz**.

model photosynthetic complexes [3, 9, 10, 15]. An intriguing question is to think about alternatives to 2D-ES that yield the same information about couplings. The techniques of this book suffice to show that neither a broadband nor a narrowband PP' signal alone can make such distinctions, but a phase-cycling procedure based

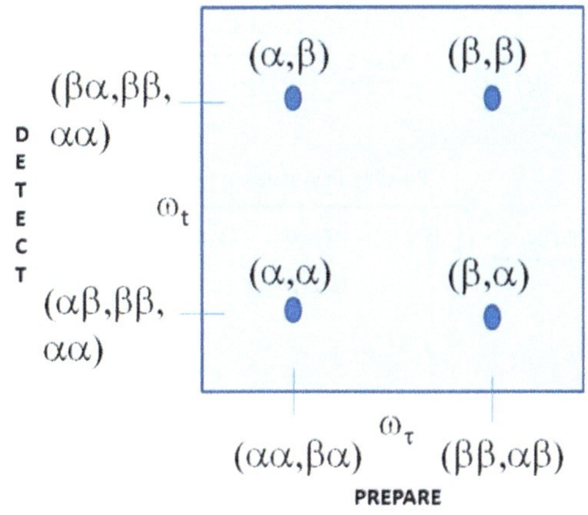

Figure E.4. Summary of QPT for a coupled dimer in the rephasing 2D-ES. The horizontal axis for the coherence frequency ω_τ is associated with a state preparation, whereas the vertical axis for the echo frequency ω_t corresponds to a detection. The four resonances labeled as (m, n) correspond to peaks located at $(\omega_\tau, \omega_t) = (\omega_{mg}, \omega_{ng})$. Their amplitudes contain information on $\chi_{ijqp}(T)$, where $|q\rangle\langle p|$ is the state prepared at the beginning of the waiting time interval, and $|i\rangle\langle j|$ the state detected at the end of that interval. (Reprinted with permission from Yuen-Zhou J and Aspuru-Guzik A 2011 *J. Chem. Phys.* **134** 134505. Copyright 2011, American Institute of Physics.)

on multicolor PP' (see example 9) or TG (example 10) spectroscopy can generate signals that are quite similar to the cross-peaks of 2D-ES, or in fact can be adapted to generate entire 2D-ES. The reader is encouraged to develop further intuition on these observations by thinking about 2D-ES as a generalization of PP' spectroscopy.

Remember that the simple structure for $S_{mn}(T)$ in equations (E.11) and (E.12) derives from the simple model of the vibrationless coupled dimer. For arbitrary dimers with coupling to vibrations, each $S_{mn}(T)$ is, in general, a linear combination of several elements of $\chi(T)$, which can be recovered by QPT [16]. Figure E.4 shows the general pattern of initial and final states that one can detect from the amplitude in each peak of the spectrum, going beyond the ideal coupled dimer.[1] This pattern can be understood by keeping track of the different DS-FDs for a general coupled dimer, which generalize figure E.2 to arbitrary coherence and population transfers in the waiting time T [16]. For instance, the diagonal peak $(\omega_\tau, \omega_t) = (\omega_{ag}, \omega_{ag})$ traces Liouville pathways that show oscillations proportional to $e^{-i\omega_{g\alpha}\tau}e^{-i\omega_{\alpha g}t}$ or $e^{-i\omega_{g\alpha}\tau}e^{-i\omega_{f\beta}t}$. These oscillations correspond to an evolution of $|g\rangle\langle\alpha|$ in the coherence time, and of $|\alpha\rangle\langle g|$ or $|f\rangle\langle\beta|$ in the echo time. This information constrains the elements of $\chi(T)$ which can be monitored via this peak,

[1] Using only the peak amplitudes might not always yield the dynamical information in $S_{mn}(T)$ very accurately if the lineshapes are broad. Instead, a procedure to extract the amplitudes of the peaks, and hence eliminate the effect of the lineshapes, is sometimes necessary. Figure E.4 must then be understood in term of amplitudes $S_{qp}(T)$, and not just as the unprocessed 2D-ES peaks.

$$|g\rangle\langle g| \xrightarrow{\text{Pulse 1}} \underbrace{|g\rangle\langle\alpha|}_{\text{Gives } \omega_\tau = \omega_{\alpha g}} \xrightarrow{\text{Pulse 2}} \overbrace{\{|g\rangle\langle g|, |\alpha\rangle\langle\alpha|, |\beta\rangle\langle\alpha|\}}^{\text{Possible initial states } |q\rangle\langle p|}$$

$$\xrightarrow[\text{Waiting time } T]{\text{FREE EVOLUTION}} \overbrace{\left\{\begin{matrix} |g\rangle\langle g|, |\alpha\rangle\langle\alpha|, |\alpha\rangle\langle\beta| \\ |\beta\rangle\langle\beta| \end{matrix}\right\}}^{\text{Possible final states } |i\rangle\langle j|} \xrightarrow{\text{Pulse 3}} \underbrace{\left\{\begin{matrix} |\alpha\rangle\langle g| \\ |f\rangle\langle\beta| \end{matrix}\right\}}_{\text{Gives } \omega_t = \omega_{\alpha g} = \omega_{f\beta}}$$

$$\xrightarrow{\text{Pulse 4 (LO)}} \text{Measurement}.$$

Here, we correlate the coherence time state $|g\rangle\langle\alpha|$ with the possible states $|q\rangle\langle p|$ prepared at the beginning of the waiting time T via pulse 2, and the echo time states $|\alpha\rangle\langle g|$ and $|f\rangle\langle\beta|$ with the possible states $|i\rangle\langle j|$ detected at the end of the corresponding interval. The pulses prepare and detect states in Liouville space by exciting or de-exciting bras or kets according to the RWA (equation (3.25)). Hence, the diagram of figure E.4 comprehensively enumerates the elements $\chi_{ijqp}(T)$ that can be detected in the diagonal peak $(\omega_\tau, \omega_t) = (\omega_{\alpha g}, \omega_{\alpha g})$.

For the ideal coupled dimer, $\chi_{ijqp}(T) = \delta_{iq}\delta_{jp}e^{-i\omega_{qp}T}$ (example 1–5) and, according to this diagram, this peak can only report on $\chi_{gggg}(T) = \chi_{\alpha\alpha\alpha\alpha}(T) = 1$, as confirmed by our previous analysis (equations (E.8*a*)–(E.8*c*)). For a simple Markovian dissipative model where there is population transfer $|\beta\rangle$ to $|\alpha\rangle$ $(\chi_{\alpha\alpha\beta\beta}(T) > 0)$ and decoherence monotonically destroys $\chi_{\alpha\beta\alpha\beta}(T)$, we expect that the quantum beats in the cross-peaks decay with the decoherence timescale.

Figure E.4 does not show processes involving $|g\rangle$, as they are assumed to only contribute as T-independent GSB background $\chi_{gggg}(T)$, that is, no decays to the GSM are considered within the timescale, while the diagonal peaks $(\omega_{\alpha g}, \omega_{\alpha g})$ and $(\omega_{\beta g}, \omega_{\beta g})$ inform on the kinetic population processes $\chi_{\beta\beta\beta\beta}(T)$ and $\chi_{\alpha\alpha\beta\beta}(T)$ of the QPT experiment.

Bibliography

[1] Biggs J D and Cina J A 2009 Calculations of nonlinear wave-packet interferometry signals in the pump–probe limit as tests for vibrational control over electronic excitation transfer *J. Chem. Phys.* **131** 224302

[2] Biggs J D and Cina J A 2009 Using wave-packet interferometry to monitor the external vibrational control of electronic excitation transfer *J. Chem. Phys.* **131** 224101

[3] Brixner T, Stenger J, Vaswani H M, Cho M, Blankenship R E and Fleming G R 2005 Two-dimensional spectroscopy of electronic couplings in photosynthesis *Nature* **434** 625–8

[4] Cheng Y C and Fleming G R 2008 Coherence quantum beats in two-dimensional electronic spectroscopy *J. Phys. Chem.* A **112** 4254–60

[5] Cho M 2009 *Two Dimensional Optical Spectroscopy* (Boca Raton, FL: CRC Press)

[6] Cina J A 2000 Nonlinear wavepacket interferometry for polyatomic molecules *J. Chem. Phys.* **113** 9488–96

[7] Cina J A 2008 Wave-packet interferometry and molecular state reconstruction: spectroscopic adventures on the left-hand side of the Schrödinger equation *Ann. Rev. Phys. Chem.* **59** 319–42

[8] Cina J A, Kilin D S and Humble T S 2003 Wavepacket interferometry for short-time electronic energy transfer: Multidimensional optical spectroscopy in the time domain *J. Chem. Phys.* **118** 46–61

[9] Collini E, Wong C Y, Wilk K E, Curmi P M G, Brumer P and Scholes G D 2010 Coherently wired light-harvesting in photosynthetic marine algae at ambient temperature *Nature* **463** 644–8

[10] Engel G S, Calhoun T R, Read E L, Ahn T K, Mancal T, Cheng Y C, Blankenship R E and Fleming G R 2007 Evidence for wavelike energy transfer through quantum coherence in photosynthetic systems *Nature* **446** 782–6

[11] Humble T S and Cina J A 2004 Molecular state reconstruction by nonlinear wave packet interferometry *Phys. Rev. Lett.* **93** 060402

[12] Humble T S and Cina J A 2006 Nonlinear wave-packet interferometry and molecular state reconstruction in a vibrating and rotating diatomic molecule *J. Phys. Chem.* **B 110** 18879–92

[13] Jonas D M 2003 Two-dimensional femtosecond spectroscopy *Ann. Rev. Phys. Chem.* **54** 425–63

[14] Mukamel S 1995 *Principles of Nonlinear Optical Spectroscopy* (Oxford: Oxford University Press)

[15] Panitchayangkoon G, Hayes D, Fransted K A, Caram J R, Harel E, Wen J, Blankenship R E and Engel G S 2010 Long-lived quantum coherence in photosynthetic complexes at physiological temperature *Proc. Natl Acad. Sci. USA* **107** 12766–70

[16] Yuen-Zhou J and Aspuru-Guzik A 2011 Quantum process tomography of excitonic dimers from two-dimensional electronic spectroscopy. I. General theory and application to homodimers *J. Chem. Phys.* **134** 134505

Appendix F

Isotropic averaging of signals

For simplicity in the exposition, we have assumed throughout the book that all the molecules in the laser spot have the same orientation with respect to the optical field polarization e. However, many ensembles are not orientationally ordered. During the femtosecond timescale of a single shot in the experiment, the molecules can be regarded as static, so molecular reorientations (for example during the time T) in electronic or vibrational spectroscopy are not relevant. The measured signal in this case is an average over the signals produced with the molecular orientations chosen from some (static) probability distribution of orientations. The reader can check that all the conclusions about phase-matching still hold in the presence of such a distribution. One of the most important distributions is the *isotropic* one, such as the one existing in homogeneous solutions, where the molecular orientation is completely random. We shall only address this case.

Starting very generally, for linear spectroscopy (section 4.1), we are interested in the isotropic average of the product of two dipole projections with the corresponding electric fields $\langle(\boldsymbol{\mu}_a\cdot\mathbf{e}_1)(\boldsymbol{\mu}_b\cdot\mathbf{e}_2)\rangle_{\text{iso}}$, where the expectation value is taken over uniformly distributed orientations of the molecule, but $\mathbf{e}_1$ and $\mathbf{e}_2$ are fixed in the lab frame. After some algebra, one can find [1, 2],

$$\begin{aligned} \langle(\boldsymbol{\mu}_a\cdot\mathbf{e}_1)(\boldsymbol{\mu}_b\cdot\mathbf{e}_2)\rangle_{\text{iso}} &= \sum_{m_1 m_2} I^{(2)}_{e_1 e_2;m_1 m_2}(\boldsymbol{\mu}_a\cdot\mathbf{m}_1)(\boldsymbol{\mu}_b\cdot\mathbf{m}_2), \\ I^{(2)}_{e_1 e_2;m_1 m_2} &= \frac{1}{3}\delta_{e_1 e_2}\delta_{m_1 m_2}, \end{aligned} \tag{F.1}$$

where $\mathbf{e}_i$ and $\mathbf{m}_i$ are unit vectors $\mathbf{x}$, $\mathbf{y}$ or $\mathbf{z}$ along the respective directions[1], and the right-hand side corresponds to quantities in the *fixed molecular frame*. Alternatively, the elegant methods of cosine vectors, Euler angles and tensor algebra yield this result

[1] Arbitrary linear, circular or elliptical polarizations $\mathbf{e}_i$ can be dealt with by appropriate linear combination of these results.

right away. We refer the reader to the pedagogical book by Barron for more details on this procedure (see [1], chapter 4).

As a consistency check, let $\mathbf{e}_1 = \mathbf{e}_2$, as in our expressions for linear spectroscopy (section 4.1), specialize to the $\boldsymbol{\mu}_a = \boldsymbol{\mu}_b$ case, and for simplicity, take $\boldsymbol{\mu}_a$ to be real. If we define $\boldsymbol{\mu}_a \cdot \mathbf{e}_1 = \cos\theta$, it follows that $\langle(\boldsymbol{\mu}_a \cdot \mathbf{e}_1)^2\rangle_{\text{iso}} = \text{Average} \times |\boldsymbol{\mu}_a|^2$ where,

$$\text{Average} = \frac{\int_0^{2\pi} \mathrm{d}\phi \int_0^{\pi} \mathrm{d}\theta \sin\theta \cos^2\theta}{\int_0^{2\pi} \mathrm{d}\phi \int_0^{\pi} \mathrm{d}\theta \sin\theta} = \frac{1}{3}. \tag{F.2}$$

A direct application of equation (F.1) yields $\langle(\boldsymbol{\mu}_a \cdot \mathbf{e}_1)^2\rangle_{\text{iso}} = \frac{1}{3}(|\boldsymbol{\mu}_a \cdot \mathbf{x}|^2 + |\boldsymbol{\mu}_a \cdot \mathbf{y}|^2 + |\boldsymbol{\mu}_a \cdot z|^2) = \frac{1}{3}|\boldsymbol{\mu}_a|^2$, consistent with the angular average above.

Going back to the situation of linear absorption in section 4.1 and using equation (F.1), the absorption of a pulse P (with a fixed polarization) by an isotropically distributed sample is simply the average of the absorption were the pulse polarized along the **x**, **y** and **z** directions with respect to the frame of the molecule. Denoting $\langle S_P\rangle_{\text{iso}}|_{\mathbf{e}_P}$ the isotropically averaged signal for linear absorption of pulse P with polarization $\mathbf{e}_P$,

$$\begin{aligned} \langle S_P\rangle_{\text{iso}}\big|_{\mathbf{x}} &= \langle S_P\rangle_{\text{iso}}\big|_{\mathbf{y}} = \langle S_P\rangle_{\text{iso}}\big|_{\mathbf{y}} \\ &= \frac{1}{3}\left(S_P\big|_{\mathbf{x}} + S_P\big|_{\mathbf{y}} + S_P\big|_{\mathbf{z}}\right). \end{aligned} \tag{F.3}$$

Notice, importantly, that the same signal is obtained regardless of the *actual* polarization of the pulse. The same type of average is performed for frequency-resolved linear spectra $A_{\mathbf{k}_P}(\omega)$. Hence, an analytical or numerical simulation of linear spectroscopy for isotropic ensembles requires, in general, the averaging of three independent signals.

Moving on to third-order spectroscopies (PP', TG, PE, 2D-ES), the analogous equations are [1, 2],

$$\begin{aligned} \langle(\boldsymbol{\mu}_a \cdot \mathbf{e}_1)(\boldsymbol{\mu}_b \cdot \mathbf{e}_2)(\boldsymbol{\mu}_c \cdot \mathbf{e}_3)(\boldsymbol{\mu}_d \cdot \mathbf{e}_4)\rangle_{\text{iso}} &= \sum_{m_1 m_2 m_3 m_4} I^{(4)}_{e_1 e_2 e_3 e_4; m_1 m_2 m_3 m_4} \\ &\quad \times (\boldsymbol{\mu}_a \cdot \mathbf{m}_1)(\boldsymbol{\mu}_b \cdot \mathbf{m}_2)(\boldsymbol{\mu}_c \cdot \mathbf{m}_3)(\boldsymbol{\mu}_d \cdot \mathbf{m}_4), \\ I^{(4)}_{e_1 e_2 e_3 e_4; m_1 m_2 m_3 m_4} &= \frac{1}{30}\left[\delta_{e_1 e_2}\delta_{e_3 e_4} \quad \delta_{e_1 e_3}\delta_{e_2 e_4} \quad \delta_{e_1 e_4}\delta_{e_2 e_3}\right] \\ &\quad \times \begin{bmatrix} 4 & -1 & -1 \\ -1 & 4 & -1 \\ -1 & -1 & 4 \end{bmatrix} \begin{bmatrix} \delta_{m_1 m_2}\delta_{m_3 m_4} \\ \delta_{m_1 m_3}\delta_{m_2 m_4} \\ \delta_{m_1 m_4}\delta_{m_2 m_3} \end{bmatrix}: \end{aligned} \tag{F.4}$$

which are used in the same spirit as the ones for linear spectroscopy. To illustrate this usage, let us consider two types of isotropically averaged PP' signals $\langle S_{PP'}\rangle_{\text{iso}}|_{(\mathbf{e}_i)}$, the *parallel* setup:

$$\langle S_{PP'}\rangle_{\text{iso}}\big|_{(\mathbf{x},\mathbf{x},\mathbf{x},\mathbf{x})} = \langle S_{PP'}\rangle_{\text{iso}}\big|_{(\mathbf{y},\mathbf{y},\mathbf{y},\mathbf{y})} = \langle S_{PP'}\rangle_{\text{iso}}\big|_{(\mathbf{z},\mathbf{z},\mathbf{z},\mathbf{z})}$$
$$= \frac{1}{5}\sum_{(\mathbf{e}_i)\in A} S_{\text{TG}}\big|_{(\mathbf{e}_i)} + \frac{1}{15}\sum_{(\mathbf{e}_i)\in B,C,D} S_{\text{TG}}\big|_{(\mathbf{e}_i)}, \tag{F.5}$$

and the *cross-polarized* setup,

$$\langle S_{PP'}\rangle_{\text{iso}}\big|_{(\mathbf{x},\mathbf{x},\mathbf{y},\mathbf{y})} = \langle S_{PP'}\rangle_{\text{iso}}\big|_{(\mathbf{x},\mathbf{x},\mathbf{z},\mathbf{z})} = \langle S_{PP'}\rangle_{\text{iso}}\big|_{(\mathbf{y},\mathbf{y},\mathbf{x},\mathbf{x})}$$
$$= \langle S_{PP'}\rangle_{\text{iso}}\big|_{(\mathbf{z},\mathbf{z},\mathbf{x},\mathbf{x})} = \langle S_{PP'}\rangle_{\text{iso}}\big|_{(\mathbf{y},\mathbf{y},\mathbf{z},\mathbf{z})} = \langle S_{PP'}\rangle_{\text{iso}}\big|_{(\mathbf{z},\mathbf{z},\mathbf{y},\mathbf{y})}$$
$$= \frac{1}{15}\sum_{(\mathbf{e}_i)\in A} S_{PP'}\big|_{(\mathbf{e}_i)} + \frac{2}{15}\sum_{(\mathbf{e}_i)\in B} S_{PP'}\big|_{(\mathbf{e}_i)} - \frac{1}{30}\sum_{(\mathbf{e}_i)\in C,D} S_{PP'}\big|_{(\mathbf{e}_i)}, \tag{F.6}$$

where

$$\begin{aligned}
A &= \{(\mathbf{x},\mathbf{x},\mathbf{x},\mathbf{x}),(\mathbf{y},\mathbf{y},\mathbf{y},\mathbf{y}),(\mathbf{z},\mathbf{z},\mathbf{z},\mathbf{z})\},\\
B &= \{(\mathbf{x},\mathbf{x},\mathbf{y},\mathbf{y}),(\mathbf{x},\mathbf{x},\mathbf{z},\mathbf{z}),(\mathbf{y},\mathbf{y},\mathbf{x},\mathbf{x}),(\mathbf{z},\mathbf{z},\mathbf{x},\mathbf{x}),(\mathbf{y},\mathbf{y},\mathbf{z},\mathbf{z}),(\mathbf{z},\mathbf{z},\mathbf{y},\mathbf{y})\},\\
C &= \{(\mathbf{x},\mathbf{y},\mathbf{x},\mathbf{y}),(\mathbf{x},\mathbf{z},\mathbf{x},\mathbf{z}),(\mathbf{y},\mathbf{x},\mathbf{y},\mathbf{x}),(\mathbf{z},\mathbf{x},\mathbf{z},\mathbf{x}),(\mathbf{y},\mathbf{z},\mathbf{y},\mathbf{z}),(\mathbf{z},\mathbf{y},\mathbf{z},\mathbf{y})\},\\
D &= \{(\mathbf{x},\mathbf{y},\mathbf{y},\mathbf{x}),(\mathbf{x},\mathbf{z},\mathbf{z},\mathbf{x}),(\mathbf{y},\mathbf{x},\mathbf{x},\mathbf{y}),(\mathbf{z},\mathbf{x},\mathbf{x},\mathbf{z}),(\mathbf{y},\mathbf{z},\mathbf{z},\mathbf{y}),(\mathbf{z},\mathbf{y},\mathbf{y},\mathbf{z})\},
\end{aligned} \tag{F.7}$$

which follow from direct application of equation (F.4). In standard PP' spectroscopy, the *actual* first two and the last two interactions are due to the same pulse P or P', respectively. Yet, the various terms in the sums of equations (F.5) and (F.6) due to the C and D groups correspond to different polarizations for each pulse, such as (**x**, **y**, **x**, **y**), which denotes the *hypothetical* experiment where the pump pulse P is polarized along **x** for the first interaction and along **y** for the second interaction, and the probe pulse P' is along **x** for the third interaction, and along **y** when it acts as the LO.

For other types of third-order spectroscopy such as TG, PE or 2D-ES, where one can vary the actual polarizations of the four different pulses, one can imagine collecting four different types of isotropically averaged signals. Focusing on TG, for instance, we can compactly rewrite equations (F.5) and (F.6) in matrix form as,

$$\begin{bmatrix} \langle S_{\text{TG}}\rangle_{\text{iso}}\big|_{(\mathbf{e}_i)\in A} \\ \langle S_{\text{TG}}\rangle_{\text{iso}}\big|_{(\mathbf{e}_i)\in B} \\ \langle S_{\text{TG}}\rangle_{\text{iso}}\big|_{(\mathbf{e}_i)\in C} \\ \langle S_{\text{TG}}\rangle_{\text{iso}}\big|_{(\mathbf{e}_i)\in D} \end{bmatrix} = \frac{1}{30}\begin{bmatrix} 6 & 2 & 2 & 2 \\ 2 & 4 & -1 & -1 \\ 2 & -1 & 4 & -1 \\ 2 & -1 & -1 & 4 \end{bmatrix} \begin{bmatrix} \sum_{(\mathbf{e}_i)\in A} S_{\text{TG}}\big|_{(\mathbf{e}_i)} \\ \sum_{(\mathbf{e}_i)\in B} S_{\text{TG}}\big|_{(\mathbf{e}_i)} \\ \sum_{(\mathbf{e}_i)\in C} S_{\text{TG}}\big|_{(\mathbf{e}_i)} \\ \sum_{(\mathbf{e}_i)\in D} S_{\text{TG}}\big|_{(\mathbf{e}_i)} \end{bmatrix}. \tag{F.8}$$

The matrix in equation (F.8) does not have full rank (the first row is equal to the sum of the other rows), having only three linearly independent rows. There is nothing special about TG in this result other than the fact that it is a third-order spectroscopy, so we have reached an important conclusion: for third-order spectroscopies, there at most only three linearly-independent isotropically averaged signals. Yet, as shown in [3], under certain circumstances, it is possible to carry out QPT by only using polarization control, exploiting the linear independence of parallel and cross-polarized isotropically averaged signals. In particular, [3] shows how to carry out partial QPT for a homodimer by linear combination of two 2D-ES stemming from these different polarization setups. Even though our presentation of QPT in this book achieved selectivity of preparation and detection of states using pulse colors, one should not forget the availability of pulse polarizations to achieve this selectivity.

Example 14. Isotropically averaged *PP′* signal for a vibrationless homodimer

Consider the vibrationless coupled dimer (electronic system not coupled to vibrations, as in section 3.2). The *PP′* signal $S_{PP'}(T)$ for such a system is given by equations (4.44*a*)–(4.44*c*). To make things simple, specialize to the case of a homodimer (of two identical monomers), where the transition dipole moments attain a very simple form. Also, assume the broadband limit in the sense that $\lim_{\sigma_n\to 0}\tilde{\varepsilon}_n(\omega)=\eta$ (equation (4.54)) but the RWA is still applicable, so that the mentioned equations still apply. Compute $\langle S_{PP'}\rangle_{\text{iso}}\big|_{(\mathbf{x},\mathbf{x},\mathbf{x},\mathbf{x})}$ and $\langle S_{PP'}\rangle_{\text{iso}}\big|_{(\mathbf{x},\mathbf{x},\mathbf{y},\mathbf{y})}$.

Solution

We use the notation of section 3.2. In the homodimer, the site energies are the same $\omega_a=\omega_b$ so the mixing angle is $\theta=\frac{\pi}{4}$ for $J>0$.

$$\begin{bmatrix}\boldsymbol{\mu}_{\alpha g}\\ \boldsymbol{\mu}_{\beta g}\end{bmatrix}=\frac{1}{\sqrt{2}}\begin{bmatrix}-1 & 1\\ 1 & 1\end{bmatrix}\begin{bmatrix}\boldsymbol{\mu}_{ag}\\ \boldsymbol{\mu}_{bg}\end{bmatrix}, \tag{F.9}$$

$$\begin{bmatrix}\boldsymbol{\mu}_{f\alpha}\\ \boldsymbol{\mu}_{f\beta}\end{bmatrix}=\frac{1}{\sqrt{2}}\begin{bmatrix}1 & -1\\ 1 & 1\end{bmatrix}\begin{bmatrix}\boldsymbol{\mu}_{ag}\\ \boldsymbol{\mu}_{bg}\end{bmatrix}, \tag{F.10}$$

so regardless of the initial angle between $\boldsymbol{\mu}_{ag}$ and $\boldsymbol{\mu}_{bg}$, we have $\boldsymbol{\mu}_{\alpha g}=-\boldsymbol{\mu}_{f\alpha}$, $\boldsymbol{\mu}_{\beta g}=\boldsymbol{\mu}_{f\beta}$, and $\boldsymbol{\mu}_{\alpha g}$ is perpendicular to $\boldsymbol{\mu}_{\beta g}$ so, without loss of generality, we can place one of the dipoles along $\mathbf{x}$ and the other along $\mathbf{y}$, as in figure F1.

Applying equations (F.5) and (F.6) to the expressions in equations (4.44*a*)–(4.44*c*), for the parallel polarization setup,

$$\langle S_{\text{ESA}}(T)\rangle_{\text{iso}}\big|_{(\mathbf{x},\mathbf{x},\mathbf{x},\mathbf{x})}=\eta^4\Bigg[\frac{1}{5}\left(|\boldsymbol{\mu}_{\alpha g}|^4+|\boldsymbol{\mu}_{\beta g}|^4\right) -\frac{1}{15}|\boldsymbol{\mu}_{\alpha g}|^2|\boldsymbol{\mu}_{\beta g}|^2\left(\mathrm{e}^{-\mathrm{i}\omega_{\alpha\beta}T}+\mathrm{e}^{-\mathrm{i}\omega_{\beta\alpha}T}\right)\Bigg], \tag{F.11a}$$

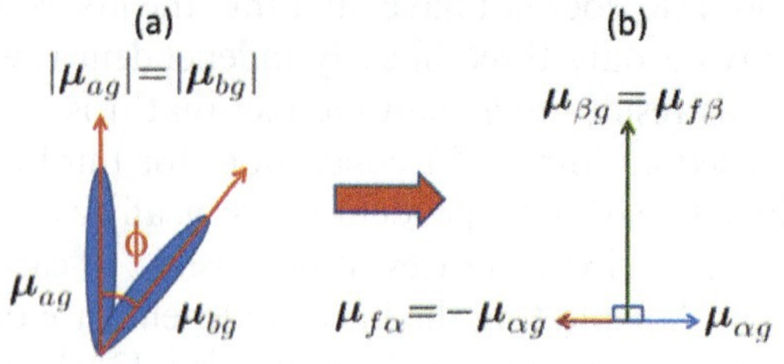

Figure F.1. Transition dipole moments for a homodimer. (*a*) The transition dipoles from the ground state g to the sites a and b are equal in magnitude but can have an angle of ϕ between them. (*b*) The transition dipoles from g to the delocalized states $|\alpha\rangle$ and $|\beta\rangle$ are perpendicular to one another, and bear simple relationships to the dipoles between α and β and f. (Adapted from Yuen-Zhou J and Aspuru-Guzik A 2011 *J. Chem. Phys.* **134** 134505. Copyright 2011, American Institute of Physics.)

$$\langle S_{\mathrm{SE}}(T)\rangle_{\mathrm{iso}}\big|_{(\mathbf{x},\mathbf{x},\mathbf{x},\mathbf{x})} = -\eta^4\left[\frac{1}{5}(|\boldsymbol{\mu}_{\alpha g}|^4 + |\boldsymbol{\mu}_{\beta g}|^4) + \frac{1}{15}|\boldsymbol{\mu}_{\alpha g}|^2|\boldsymbol{\mu}_{\beta g}|^2(\mathrm{e}^{-\mathrm{i}\omega_{\alpha\beta}T} + \mathrm{e}^{-\mathrm{i}\omega_{\beta\alpha}T})\right], \tag{F.11b}$$

$$\langle S_{\mathrm{GSB}}(T)\rangle_{\mathrm{iso}}\big|_{(\mathbf{x},\mathbf{x},\mathbf{x},\mathbf{x})} = -\eta^4\left[\frac{1}{5}(|\boldsymbol{\mu}_{\alpha g}|^4 + |\boldsymbol{\mu}_{\beta g}|^4) + \frac{2}{15}|\boldsymbol{\mu}_{\alpha g}|^2|\boldsymbol{\mu}_{\beta g}|^2\right], \tag{F.11c}$$

giving

$$\langle S_{PP'}(T)\rangle_{\mathrm{iso}}\big|_{(\mathbf{x},\mathbf{x},\mathbf{x},\mathbf{x})} = -\eta^4\left[\frac{1}{5}(|\boldsymbol{\mu}_{\alpha g}|^4 + |\boldsymbol{\mu}_{\beta g}|^4) + \frac{2}{15}|\boldsymbol{\mu}_{\alpha g}|^2|\boldsymbol{\mu}_{\beta g}|^2(1 + 2\cos\omega_{\alpha\beta}T)\right], \tag{F.12}$$

and for the cross polarization setups,

$$\langle S_{\mathrm{ESA}}(T)\rangle_{\mathrm{iso}}\big|_{(\mathbf{x},\mathbf{x},\mathbf{y},\mathbf{y})} = \eta^4\left[\frac{1}{15}(|\boldsymbol{\mu}_{\alpha g}|^4 + |\boldsymbol{\mu}_{\beta g}|^4) + \frac{1}{30}|\boldsymbol{\mu}_{\alpha g}|^2|\boldsymbol{\mu}_{\beta g}|^2(\mathrm{e}^{-\mathrm{i}\omega_{\alpha\beta}T} + \mathrm{e}^{-\mathrm{i}\omega_{\beta\alpha}T})\right], \tag{F.13a}$$

$$\langle S_{\mathrm{SE}}(T)\rangle_{\mathrm{iso}}\big|_{(\mathbf{x},\mathbf{x},\mathbf{y},\mathbf{y})} = -\eta^4\left[\frac{1}{15}(|\boldsymbol{\mu}_{\alpha g}|^4 + |\boldsymbol{\mu}_{\beta g}|^4) - \frac{1}{30}|\boldsymbol{\mu}_{\alpha g}|^2|\boldsymbol{\mu}_{\beta g}|^2(\mathrm{e}^{-\mathrm{i}\omega_{\alpha\beta}T} + \mathrm{e}^{-\mathrm{i}\omega_{\beta\alpha}T})\right], \tag{F.13b}$$

$$\langle S_{\text{GSB}}(T)\rangle_{\text{iso}}\big|_{(\mathbf{x},\mathbf{x},\mathbf{y},\mathbf{y})} = -\eta^4\left[\frac{1}{15}\left(|\boldsymbol{\mu}_{\alpha g}|^4 + |\boldsymbol{\mu}_{\beta g}|^4\right) + \frac{4}{15}|\boldsymbol{\mu}_{\alpha g}|^2|\boldsymbol{\mu}_{\beta g}|^2\right], \tag{F.13c}$$

giving

$$\langle S_{PP'}(T)\rangle_{\text{iso}}\big|_{(\mathbf{x},\mathbf{x},\mathbf{y},\mathbf{y})} = -\eta^4\left[\frac{1}{15}\left(|\boldsymbol{\mu}_{\alpha g}|^4 + |\boldsymbol{\mu}_{\beta g}|^4\right) + \frac{2}{15}|\boldsymbol{\mu}_{\alpha g}|^2|\boldsymbol{\mu}_{\beta g}|^2(2 - \cos\omega_{\alpha\beta}T)\right]. \tag{F.14}$$

The coherences and populations are weighted in different ways in the parallel and cross-polarized signals, so in principle, one can isolate these terms by linear combinations of the various signals. As an example,

$$\Re\chi_{\alpha\beta\alpha\beta}(T) = \cos\omega_{\alpha\beta}T = \frac{3}{2\eta^4|\boldsymbol{\mu}_{\alpha g}|^2|\boldsymbol{\mu}_{\beta g}|^2}\left(\langle S_{PP'}(T)\rangle_{\text{iso}}\big|_{(\mathbf{x},\mathbf{x},\mathbf{x},\mathbf{x})} - 3\langle S_{PP'}(T)\rangle_{\text{iso}}\big|_{(\mathbf{x},\mathbf{x},\mathbf{y},\mathbf{y})}\right) + 1 \tag{F.15}$$

This is the essence of the QPT procedure advocated in [3], which generalizes this example to homodimers with coupling to vibrations.

Bibliography

[1] Barron L 2004 *Molecular Light Scattering* (Cambridge: Cambridge University Press)

[2] Craig D P and Thirunamachandran T 1998 *Molecular Quantum Electrodynamics: An Introduction to Radiation Molecule Interactions* (New York: Dover Publications)

[3] Yuen-Zhou J and Aspuru-Guzik A 2011 Quantum process tomography of excitonic dimers from two-dimensional electronic spectroscopy. I. General theory and application to homodimers *J. Chem. Phys.* **134** 134505

Lightning Source UK Ltd.
Milton Keynes UK
UKOW06n1431201014

240323UK00002B/13/P

9 780750 310635